Advanced Series in Agricultural Sciences 19

Co-ordinating Editor: B. Yaron, Bet-Dagan

Editors: G.W. Thomas, Lexington
L.D. Van Vleck, Ithaca

Advanced Series in Agricultural Sciences

Volumes already published in the series

Volume 1: *A. P. A. Vink*
Land Use in Advancing Agriculture

Volume 2: *H. Wheeler*
Plant Pathogenesis

Volume 3: *R. A. Robinson*
Plant Pathosystems

Volume 4: *H. C. Coppel, J. W. Mertins*
Biological Insect Pest Suppression

Volume 5: *J. J. Hanan, W. D. Holley, K. L. Goldsberry*
Greenhouse Management

Volume 6: *J. E. Vanderplank*
Genetic and Molecular Basis of Plant Pathogenesis

Volume 7: *J. K. Matsushima*
Feeding Beef Cattle

Volume 8: *R. J. Hanks, G. L. Ashcroft*
Applied Soil Physics

Volume 9: *J. Palti*
Cultural Practices and Infectious Crop Diseases

Volume 10: *E. Bresler, B. L. McNeal, D. L. Carter*
Saline and Sodic Soils

Volume 11: *J. R. Parks*
A Theory of Feeding and Growth of Animals

Volume 12: *J. Hagin, B. Tucker*
Fertilization of Dryland and Irrigated Soils

Volume 13: *A. J. Koolen, H. Kuipers*
Agricultural Soil Mechanics

Volume 14: *G. Stanhill*
Energy and Agriculture

Volume 15: *E. A. Curl, B. Truelove*
The Rhizosphere

Volume 16: *D. P. Doolittle*
Population Genetics: Basic Principles

Volume 17: *A. Feigin, I. Ravina, J. Shalhevet*
Irrigation with Treated Sewage Effluent

Volume 18: *D. Gianola, K. Hammond*
Advances in Statistical Methods for Genetic Improvement of Livestock

Volume 19: *I. Rosenthal*
Electromagnetic Radiations in Food Science

I. Rosenthal

Electromagnetic Radiations in Food Science

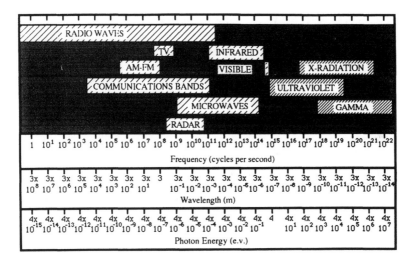

With 28 Figures and 26 Tables

Springer-Verlag
Berlin Heidelberg New York
London Paris Tokyo
Hong Kong Barcelona
Budapest

Ionel Rosenthal, Ph.D.
Department of Food Science
Agricultural Research Organization, The Volcani Center
P.O. Box 6, Bet Dagan 50250, Israel

ISBN-13: 978-3-642-77108-8   e-ISBN-3: 978-3-642-77106-4
DOI: 10.1007/978-3-642-77106-4

Library of Congress Cataloging-in-Publication Data
Rosenthal, Ionel. Electromagnetic radiations in food science/I. Rosenthal. — (Advanced series in agricultural sciences; 19) Includes bibliographical references and index. ISBN 3-540-54833-5 (Berlin) — ISBN 0-387-54833-5 (New York) 1. Radiation preservation of food. 2. Electromagnetic waves. I. Title. II. Series. TP371.8.R67   1992   664'.0288—dc20

This work is subject to copyright. All rights are reserved, whether the whole or part of the material is concerned, specifically the rights of translation, reprinting, reuse of illustrations, recitation, broadcasting, reproduction on microfilms or in other ways, and storage in data banks. Duplication of this publication or parts thereof is only permitted under the provisions of the German Copyright Law of September 9, 1965, in its current version, and a copyright fee must always be paid.

© Springer-Verlag Berlin Heidelberg 1992
Softcover reprint of the hardcover 1st edition 1992

The use of registered names, trademarks, etc. in this publication does not imply, even in the absence of a specific statement, that such names are exempt from the relevant protective laws and regulations and therefore free for general use.

Typesetting: Best-Set Typesetter Ltd., Hong Kong

31/3145-5 4 3 2 1 0 – Printed on acid-free paper

Preface

This book has been written for those whose interests bridge food processing and physicochemical aspects of radiation. It is not intended to be a comprehensive review of publications concerning foods and radiations. Instead, it is an attempt to familiarize the reader with pertinent knowledge of a unified, interdisciplinary concept of various electromagnetic radiations and corresponding effects on foods. Consideration was given to similarities and differences between various segments of the electromagnetic spectrum. The broad approach of this book was considered to be crucial for cross-discipline comparisons.

The reader is introduced to the electromagnetic spectrum in the Prologue and then the book follows the wavelengths, from short to long values. Chapter 1 deals with ionizing radiation: historical background, sources of radiation employed in food treatment, units of measurement, and fundamentals of radiation chemistry. A survey of potential applications of ionizing radiation in food technology is followed by a description of methods for radiation dosimetry. Safety and wholesomeness of irradiated foods, analytical methods for postirradiation dosimetry in foods, and consumer acceptance of food irradiation conclude this section. Chapter 2 intrudes into the next segment of the spectrum: ultraviolet-visible radiation. The general presentation of this electromagnetic emission and illumination source enables the discussion of its effects on foods, including applications in food analysis. Chapter 3 covers infrared heating and analytical applications of infrared radiation. Microwave radiation is the topic of Chapter 4: molecular mechanisms of heating with microwaves, equipment and applications in the food industry are discussed. Chapter 5 covers two case studies which exemplify the practical potential of ionizing radiation in food treatment: suppression of postharvest pathogens of fresh fruits and vegetables, and decontamination of poultry meat. Both of these topics deal with the interaction between microorganisms and food.

It goes without saying that a book like this cannot be written on the exclusive basis of personal observations. Reading what others have written has been a salient source of information and the

VI   Preface

greatest debt is owed to the many researchers whose detailed observations made this volume possible. References are gathered at the end of the book. Unreviewed and less widely available publications have been avoided as much as possible.

The content of the book recalls the many pleasant associations which the author has had with people and research projects in the fields of radiations and food chemistry. The contributions by Drs. Moshe Faraggi, Rivka Barkai-Golan, I. Klinger and M. Lapidot are most appreciated. I also thank Mrs. S. Bernstein for so ably preparing the manuscript for the press.

In closing, let me acknowledge some of those who have contributed indirectly, though not insubstantially, to the success of this project. Haya, my partner through life, and Tal, Shirrie and Dana have shown interest, patience, and concern both through the "highs" and "lows" of this project.

Bet Dagan, Israel, October 1992   *Ionel Rosenthal*

Contents

Prologue   Electromagnetic Radiation   1

Chapter 1   Ionizing Radiation   9

1.1 Sources and Units of Measurement of Ionizing Radiation   9
     1.1.1 Sources of Ionizing Radiation   9
     1.1.2 Units of Measurement   11
1.2 Interaction of Ionizing Radiation with Matter.
     Chemistry of Radiation   12
     1.2.1 Interaction of Ionizing Radiation with Matter   12
     1.2.2 Radiation Chemistry   17
1.3 Applications of Ionizing Radiation in Food Technology ..   20
1.4 Radiation Dosimetry (*M. Faraggi*)   32
     1.4.1 Physical Dosimetric Methods....................   32
     1.4.2 Chemical Dosimetric Methods   33
1.5 Safety and Wholesomeness of Irradiated Foods   36
1.6 Analytical Methods for Postirradiation Dosimetry
     of Foods..   40
     1.6.1 Measurements of Physical Effects   55
     1.6.2 Measurements of Chemical Effects..............   58
     1.6.3 Microbiological and Biological Methods   61
     1.6.4 Conclusions   62
1.7 Consumer Acceptance of Food Irradiation   63

Chapter 2   Ultraviolet-Visible Radiation   65

2.1 The Definition of Ultraviolet-Visible Radiation   65
2.2 Interactions Between Ultraviolet-Visible Radiation
     and Matter..   66
     2.2.1 Absorption and Emission of Light   66
     2.2.2 Polarized Light...............................   72
2.3 Photooxidation   72
2.4 Illumination Sources and Units of Measurement   77
     2.4.1 Illumination Sources   77

VIII   Contents

2.4.2 Units of Measurement 79
2.5 Biological Effects and Safety Aspects
    of Ultraviolet-Visible Radiation...................... 79
2.6 Effects of Ultraviolet-Visible Radiation on Foods 81
    2.6.1 The Color of Foods 81
    2.6.2 Beneficial Effects of Light in Production of Foods.. 83
    2.6.3 Photodegradation of Foods 88
    2.6.4 Light Absorbers and Photoinitiators in Foods 89
    2.6.5 Constituents of Foods Sensitive
          to Photodegradation 90
2.7 Applications of Ultraviolet-Visible Radiation
    in Food Analysis.................................. 101

Chapter 3   Infrared Radiation 105

3.1 The Definition of Infrared Radiation 105
3.2 Sources of Infrared Radiation 106
3.3 Infrared Heating in Food Processing.................. 107
3.4 Analytical Applications of Infrared Radiation........... 112

Chapter 4   Microwave Radiation....................... 115

4.1 The Definition of Microwave Radiation 115
4.2 Molecular Mechanisms of Heating with Microwaves 115
4.3 Equipment for Microwave Heating 126
4.4 Applications of Microwave Heating in the Food Industry 129
4.5 Materials for Food Containers for Microwave Treatment   145
4.6 Safety Aspects of Microwave Heating Equipment 147
4.7 Analytical Applications of Microwave Radiation 149

Chapter 5   Case Studies 155

5.1 Suppression of Postharvest Pathogens of Fresh Fruits
    and Vegetables by Ionizing Radiation (*R. Barkai-Golan*)   155
    5.1.1 Introduction 155
    5.1.2 Radiation Effects on Pathogens 156
    5.1.3 Radiation Effects on Disease Development 167
    5.1.4 Pathological and Microbiological Problems
          Following Irradiation 189
    5.1.5 Conclusions 192
5.2 Decontamination of Poultry Meat by Ionizing Radiation
    (*I. Klinger and M. Lapidot*) 194
    5.2.1 Microbiological Quality of Processed Poultry Meat 194
    5.2.2 Pathogenic Bacterial Contaminants
          of Poultry Meat 196

Contents   IX

5.2.3 Decontamination of Poultry Meat
by Ionizing Radiation 201

References 209

Prologue
Electromagnetic Radiation

Electromagnetic radiation can be regarded as having a dual nature. From the time of Newton until the advent of the quantum theory, the predominant apprehension of electomagnetism was the wave theory. The propagation phenomena such as reflection, refraction, diffraction, polarization and, particularly, interference can be explained in terms of propagation of a wave. However, the actual nature of the wave and the mechanism of its propagation were not established until the latter part of the nineteenth century. In the 1860s James Clerk Maxwell made one of the major contributions to physics. He demonstrated by powerful mathematical reasoning that an oscillating magnetic field was associated with a similar electric field, if a wave was propagated in a direction perpendicular to a plane containing these fields. Maxwell's equations indicated that the velocity of propagation of an "electromagnetic" wave in vacuo is numerically identical to the velocity of light. In 1887 Hertz confirmed Maxwell's prediction of propagated waves from systems involving oscillating electrical and magnetic fields.

Maxwell's electromagnetic field theory describes radiation in terms of oscillating electric (E) and magnetic (H) fields operating in planes which are perpendicular to each other and to the direction of progress. The time-variable strength of the resulting vector is described as a transverse wave by a sinusoidal function (Fig. 1).

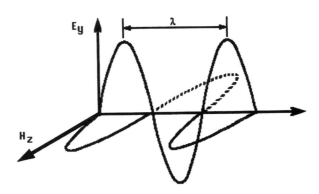

Fig. 1. A wave of electromagnetic radiation

The wave of radiation is characterized by wavelength (λ), which is the distance between identical positions on successive waves, or by frequency (v), which is the number of waves that pass a fixed point per second. These two parameters are related in the following equation:

$$v = \frac{c}{\lambda} \, ,$$

where c is the velocity of electromagnetic radiation in a vacuum (2.9979×10^8 m/s). The wavelength is measured in units of length, which extend from nanometers, nm (formally named milimicrons, $m\mu$) up to meters. The use of Å units (angstrom, 10 Å = 1 nm) is not recommended any more. The conversion factors between these units are as follows:

$$1\,m = 10^3\,mm = 10^6\,\mu m = 10^9\,nm.$$

Electromagnetic radiation is a field which propagates energy at a fixed velocity. The velocity of propagation in a medium, c', depends upon both ε, the dielectric constant of the medium, and μ, the magnetic permeability:

$$c' = c/(\varepsilon/\mu)^{1/2} = c/n.$$

The factor n by which the velocity of electromagnetic radiation in a vacuum is diminished in a medium is the refractive index of the medium relative to a vacuum. The refractive index is a function of wavelength. For example, for the 589 nm sodium line (frequently used in polarimetric measurements for sugar analysis), n is 1.00029 for air and 1.33 for water at 25 °C.

For propagation of radiation in the x direction, the electric field vector E, which is customarily plotted in the y direction, is a function of the wavelength, λ, and time:

$$E_y = A \sin 2\pi(x/\lambda - vt + \Phi),$$

where A is the amplitude of the electric vector. The magnetic vector, H, is at right angles to the electric vector, E, and is given by:

$$H_z = (\varepsilon/\mu)^{1/2} A \sin 2\pi(x/\lambda - vt + \Phi),$$

where $(\varepsilon/\mu)^{1/2} A$ is the amplitude of the magnetic vector.

The term Φ is a phase factor. Natural radiation is incoherent because Φ varies for the many photons making up the beam. Coherent radiation is produced by lasers and all constituent photons have the same phase relationship.

The above equations refer to "plane-polarized" light, where $E_z = H_y = 0$. Unpolarized light has random orientations of E vectors; in circularly polarized light there is a specific fixed phase relationship between coherent E_y and E_z vectors (but E always remains perpendicular to H).

Electromagnetic Radiation

The most significant modification of Maxwell's nineteenth-century picture of electromagnetic radiation is the awareness that wave motion may be associated with particulate properties. Where there is an exchange of energy between the radiation field and matter it is necessary to treat the field as a particle with momentum, rather than as a wave. Such a particle is termed a photon. According to the quantum theory, the absorption or emission of radiation must occur in discrete units called quanta or photons and the energy of the particle, E, and frequency of the wave, v, are related:

$$E = hv = \frac{hc}{\lambda} \, ,$$

where h is a universal constant called Planck's constant ($6.6256 \times 10^{-34}\,\mathrm{J\,s}$). It follows that the energy carried by an electromagnetic radiation is inversely proportional to the wavelength.

The electromagnetic spectrum covers a wide range of frequencies and corresponding energies (Table 1). The range spreads over a difference in magnitude of 10^{15}, from gamma rays of 10^{21} Hz frequency and wavelengths in the order of 10^{-13} m, to radio waves of 10^6 Hz frequency and wavelengths longer than 1 m. The energies carried by different electromagnetic

Table 1. The spectral distribution and sources of electromagnetic radiation

Radiation	Wavelength	Frequency (Hz)	Energy (eV)	Source
Gamma rays	0.0005–0.01 nm	6×10^{20}–3×10^{19}	2.5×10^6–1.2×10^5	Radioactive decay
X-rays, accelerated electrons	0.01–10 nm	3×10^{19}–3×10^{16}	1.2×10^5–1.2×10^2	Machine generated
Vacuum ultraviolet	10–200 nm	3×10^{16}–10^{15}	1.2×10^2–4.1	Mercury or deuterium lamps
Ultraviolet	200–400 nm	10^{15}–7×10^{14}	4.1–2.9	
Visible	400–800 nm	7×10^{14}–4×10^{14}	2.9–1.5	Incandescent or fluorescent lamps
Near infrared	0.8–1.5 μm	4×10^{14}–2×10^{14}	1.5–0.8	Incandescent bodies
Infrared	1.5–5.6 μm	2×10^{14}–5×10^{13}	0.8–0.2	
Far infrared	5.6–1000 μm	5×10^{13}–3×10^{11}	0.2–10^{-3}	
Microwaves	1–1000 mm	3×10^{11}–3×10^8	10^{-3}–10^{-6}	Magnetron or klystron
TV and radio waves	1–550 m	3×10^8–5×10^5	10^{-6}–2×10^{-9}	Piezoelectric, ferroelectric or magneto-strictive-transducers

radiations vary over the same range of 10^{15}. Thus, at the high energy end of the spectrum are gamma rays, produced during changes inside the atomic nucleus and which carry more than 1 million eV. Then come X-rays with energies in the hundred thousand electron-volt range and with wavelengths in the order of 10^{-10} m. The X-rays are produced by energy state transitions of inner electrons close to the nucleus. Then, overlapping the low energy side of X-rays, is the ultraviolet radiation generated by transitions of the outer or valence electrons and this, in turn, merges into the visible region. The energies associated with electronic transitions in the ultraviolet-visible region have fallen to approximately 10 eV with corresponding wavelengths of 10^{-7} m. Then comes the infrared segment which has an energy of a fraction of 1 eV, equivalent to molecular interactomic bonding. The infrared region blends into the microwave region where also intra- and intermolecular forces and interactions are involved. Finally, the electromagnetic spectrum is occupied by TV and radio waves with wavelengths measured in meters. There is no one universal source for practical generation of all electromagnetic radiations; for each type of radiation there is a specific piece of equipment. Furthermore, in spite of the theoretical equivalence of energy–wavelength–frequency, in practice, there is a preferred way of expression for each type of radiation. Thus, the ionizing radiation is customarily defined in energy terms, the range from ultraviolet to infrared in wavelengh units, and microwaves in frequency units.

Not every spatial oscillation that propagates as a wave is electromagnetic. For example, sonic vibrations are characterized by frequency values in the low range of the electromagnetic spectrum but the wave does not possess electric or magnetic components. Sonics are only the result of changes in the kinetic energy of gas giving rise to pressure fluctuations which propagate by oscillations. Since the velocity of sound in air is in the order of 300 m/s, roughly one million times smaller than the velocity of light, wavelengths between 1 mm and 1 m of electromagnetic radiation correspond in acoustics to frequencies between 300 Hz and 300 kHz. Therefore, an apparent similarity exists between microwaves and acoustics. It should be noted that the upper limit for the waves within the range audible to the average human is 20 kHz.

The emission of radiant energy is the result of intra- or interatomic movements within the radiation source which, in roughly quantic terms, are descending transitions between different energy levels. It is not surprising that absorption of this energy, after traveling through an inert medium, can induce ascending transitions between similar energy levels within another molecule ready to accept it (Fig. 2).

In any system containing electrons or nucleons there will be a certain configuration which has minimum energy, named the "ground state". If the configuration of the molecule changes from that of the ground state, then the new configuration will have an energy higher than that of the ground state. There will, of course, be many different excited states, corresponding to different possible configurations of the electrons or nucleons. In passing from the ground state to the excited state the system needs extra energy.

The interaction of radiation with matter involves this exchange of energy. Exposing a molecule to radiation with the energy or wavelength equal in value to the difference in energy between two levels for that molecule, will alter the energy within the molecule, and a photon will be annihilated. This event is termed the absorption of radiation. Modern theoretical models predict precisely the range of values for the exchanged energy by atoms and molecules. Absorption of radiation creates an excited state and annihilates a photon.

In theory, all the energy levels have definite and precise values, and the transitions have zero frequency spread. In practice, every transition has a

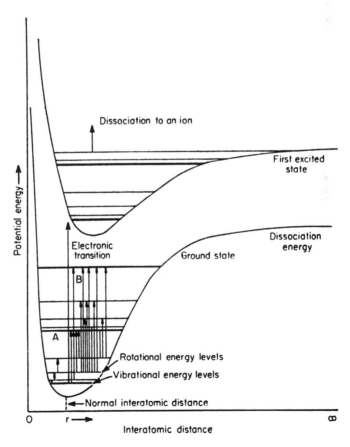

Fig. 2. Potential energy diagram of rotational, vibrational and electronic transitions

6 Electromagnetic Radiation

finite width which arises from an indefiniteness in the actual energy value of the different energy levels and can be caused by various factors, such as thermal vibrations, collisions with other molecules, or interactions with the magnetic or electric fields of surrounding atoms or nuclei.

The excited state is a distinct chemical species. It is different from the parent or unexcited ground state that absorbed radiation. The excited state has excess energy derived from the absorbed photon, and this energy will ultimately be dissipated. As a result of this energy input, the absorbing material is prone to molecular changes which are related directly to the amount of energy absorbed. There are three pathways for the dissipation of this excess energy: (1) creation of molecular motion – thermal relaxation; (2) emission of radiation – luminescence; (3) reorganization of the molecule's architecture, that is, chemical modifications. Hence, the lifetime of an excited state is governed by these three processes. Each of these channels of decay events proceeds at a given rate. The excited state is therefore only a very transient chemical species produced after an absorption event.

These radiation-induced molecular changes and their applications in food chemistry are summarized in Table 2.

Many segments of the electromagnetic spectrum have applications in food science and technology. A priori, the advantage of activating a process

Table 2. The origin, effects and applications of electromagnetic energy

Radiation	Molecular origin	Primary molecular effect	Application in food technology
Gamma rays	Nuclear transitions	Cleavage of chemical bonds, ionization	In-depth bactericidal action, enzyme inactivation
X-rays	Atomic inner shell electrons	Cleavage of chemical bonds, ionization	Analytical chemistry
Vacuum ultraviolet	Atomic outer shell electrons	Ionization, electronic excitation	
Ultraviolet		Electronic excitation	Superficial bactericidal action, food analysis
Visible		Electronic excitation	Food analysis
Near infrared	Vibrations and rotations of chemical bonds	Vibrations and rotations of chemical bonds	
Infrared			Food analysis
Far infrared			Superficial heating
Microwaves	Oscillations of mobile or free electrons	Molecular polarization	In-depth heating
TV and radio waves		Spin orientation in a magnetic field	Food analysis

Electromagnetic Radiation

with electromagnetic energy is the specificity of the response, as the same kind of radiation may interact differently with the components of a mixture. This means that side effects can be minimized and the time savings in achieving a desired result may be substantial. Sometimes application of radiation procedures, such as infrared and microwave heating, or preservation of food by ionizing radiation, can be superior to alternative treatments. Under other circumstances the radiation action could be most beneficial as being complementary to conventional methods.

In some applications, like spectroscopic analysis of foods, the role of radiation is unique and has no substitute. Spectrophotometric examination involves application of a well-characterized energy on the sample to be analyzed. This energy may be in the form of X-rays, ultraviolet, visible, infrared, or microwaves. The interaction of the incident energy with the sample occurs in some unique manner dictated by the physical or chemical properties of the sample and, consequently, it may be reflected, transmitted, or absorbed. Measuring and recording the difference between energy input and output provides a correlation with product quality. The advances in analytical spectroscopy are spread over virtually the entire electromagnetic range and provide techniques which are indispensable in modern food quality control.

Radiant energy is a natural phenomenon and during the millennia of evolution living matter on earth has adapted to the existing levels. However, for industrial applications the doses needed might be much higher than found in nature. For this reason, along with the increasing use of radiations, a lot of attention has been focused on possible biological effects. Absorption of radiation by atoms of living matter induces primary physicochemical changes which occur in a very small fraction of a second. Sub-

Table 3. Primary effects of electromagnetic radiation

Type of radiation	Possible primary effects	Factors affecting the effect
Gamma and X-rays, electron beam	Unselective ionization, free radical reactions	Radiation dose, temperature, presence of oxygen, composition of the exposed material (water content)
Ultraviolet	Electronic excitation of nucleic acids and proteins, photosensitization	Incident power and irradiation dose and rate
Visible	Electronic excitation of natural pigments, photosensitization	Presence of oxygen and pigments, irradiation dose and rate
Infrared	Heating	Incident power and dose
Microwave	Heating	Electric field strength, dielectric properties

sequent processes, whereby these changes are expressed in terms of damage to life, may take hours, months, or even decades. The effects are drastically dependent on the wavelength (Table 3). The potential for permanent biological injury is highest from ionizing radiation (gamma rays, X-rays, electron beams) because of the absence of proper, natural protective means.

Chapter 1   Ionizing Radiation

1.1 Sources and Units of Measurement of Ionizing Radiation

1.1.1 Sources of Ionizing Radiation

The high-energy radiations which are of primary interest for food preservation are gamma rays, high-speed electron beams and X-rays. All
these radiations are characterized by deep penetrating power and contain
enough energy to break chemical bonds and ionize molecules in their path,
without appreciably raising the temperature. A commercial source of
ionizing radiation is characterized by the nature of radiation emitted, its
energy distribution, and the rate of emission.

Gamma emission is produced by the nuclear disintegration of certain
radioactive materials. A radioactive transformation is the change of an
atom from one element to another by the involvement of particulate
radiation (alpha, beta, neutrons). The gamma radiation released during
some radioactive transformations is probably due to the transition of the
daughter nucleus from a higher energy level (excited state) to the ground
state and it consists of electromagnetic radiation of sharply defined wavelengths. The least expensive sources of radiation for food preservation are
gamma rays from the nuclides cobalt-60 and cesium-137. These radioactive
elements are either byproducts of atomic fission or waste products of the
atomic industry. ^{60}Co is a radioactive isotope of cobalt, prepared artificially
by bombarding natural cobalt (^{59}Co) with neutrons in a nuclear reactor.
^{60}Co has a half-life of 5.3 years and emits gamma rays of 1.17 and
1.33 MeV. ^{137}Cs is a radioactive isotope of cesium which is obtained as a
fission product from uranium and other elements in a nucler reactor. ^{137}Cs
has a half-life of 30 years and emits gamma rays with an energy of
0.66 MeV. The use of ^{60}Co is much more common since this isotope is more
easily available and safer to use. The supply of ^{137}Cs seems to be more
limited because of regulatory restrictions imposed on the processing of
spent reactor fuel. Since radioactive cesium is actually manufactured as its
chloride salt, there is also a risk associated with the use of a water-soluble
compound in the event of an accidental leakage through the protective
stainless steel envelope.

Electron beams are machine-produced. They can be emitted as cathode
rays from the cathode of an evacuated tube subjected to an electrical

potential or are produced in linear accelerators. For food treatment the energy of electron beams is limited to 10 MeV, at most.

X-rays are the result of bombarding heavy metal targets (tungsten, molybdenum) with high-velocity electrons within an evacuated tube. When fast electrons hit matter they lose energy by collision with electrons of the target material and by deflection by nuclei. In the collision process the electrons of the target material are excited or even completely ejected from the atom, at the expense of some energy from the incident high-energy electron. The emitting radiation appears as electromagnetic radiation of a wide and almost continuous range of wavelengths superimposed on a few "characteristic" lines. Since each element has distinctive energy levels, these "characteristic" X-rays are specific for the element bombarded. The fact that X-rays consist of a wide variety of wavelengths is the main difference from gamma radiation.

At the present time the use of X-rays for the treatment of food on an industrial scale is not proposed, but the principle involved is the same as that in the employment of gamma radiation. For food treatment the X-ray machines should be operated at an energy level of 5 MeV, or lower.

Fast electrons have much less penetrating power than X-rays or gamma radiation of the same energy. The penetration of electrons is determined by the energy of the rays and the density of the target material. The interaction between electrons (or beta particles as they are called if they are produced in radioactive decay processes) and matter involves exactly the same processes of excitation and ionization as for gamma rays. However, because of the small mass and single negative charge, each time the electron approaches the target atoms it is deflected from its path by orbital electrons and by positive atomic nuclei. For these reasons electrons have a poorer penetrability compared to gamma rays. For example, gamma radiation with an energy between about 0.15 and 4 MeV will penetrate about 30 cm of water. In comparison, accelerated electrons with an energy of 10 MeV, which is the maximum allowed for food applications, will penetrate only to a depth of about 4 cm. Therefore, electrons are used for products of limited thickness or for superficial treatment, although double-sided irradiation improves penetration. Except for the penetration power, treatment of food with fast electrons can be regarded, in all other respects, as equivalent to treatment with a similar dose of gamma radiation.

The restrictions to energy levels of 5 MeV for X-rays and 10 MeV for electron beams are explained by the need to prevent nuclear changes which might create artificial radioactive elements in food. Tangentially, it is noted that because of the significant potential for inducing radioactivity, neutron beams, which exhibit a great penetration power and are very effective in the destruction of bacteria, are inappropriate for use with food.

The heart of a food irradiation plant is the radiation source. A thick protective wall of concrete, usually over 1.7 m thick, surrounds the plant to

Units of Measurement 11

contain the radiation. An irradiation chamber with radionuclides (which are encapsulated in rod-shaped stainless steel holders to ensure insulation from the external environment) also contains a 5–6 m deep water pool for storage of the radiation source when not in use. The apertures of the irradiation chamber for introduction and removal of products must not allow accidental leakage of ionizing radiation. Since, like any other electromagnetic radiation, the ionizing radiation travels in a straight line, the passages to and from the irradiation chamber are usually in the form of labyrinths, with at least three right-angle bends. An array of safety latches prevents accidental intrusion of personnel into the operational area. The irradiation plants can operate continously or by batch. In an efficient operation a conveyor belt usually passes food near to the source. The strength of the source, the speed of the conveyor, and the shape of the packages determine the amount of radiation received by the material treated.

1.1.2 Units of Measurement

Electron volt. The amount of kinetic energy gained by an electron accelerated through an electric potential difference of 1 V is 1 eV.

Becquerel. This is a unit of radioactive intensity of a radionuclide. It is equal to one disintegration per second.

Curie. The basic unit used to describe the intensity of radioactivity is the curie. It was originally defined as the number of disintegrations per second which occur in 1 g of pure radium. One curie equals that quantity of radioactive material having 3.7×10^{10} disintegrations per second, which means that 1 Ci is equivalent to 3.7×10^{10} Bq. This unit is not adequate for radiation work since the rate of energy release depends not only on the rate of disintegration, but also on the energy liberated per disintegration, which is different for different radionuclides.

Roentgen. This was the original unit of measurement of radiation and was defined in terms of ionization events. A roentgen of radiation has been defined as the quantity of radiation which produces 1 esu (electrostatic unit) (2.083×10^9 ion pairs) of positive or negative electricity per cm^3 of air at standard temperature and pressure. This does not give a direct measurement of the absorbed dose. It actually yields a measurement of the amount of charge produced by a given exposure to a particular energy radiation absorbed by dry air in the ion chamber. In a liquid, however, direct experimental measurement of the amount of charge produced is virtually impossible. Therefore, a number of units have been proposed for the evaluation of the radiation absorbed by a substrate. These include the "roentgen equivalent physical" (rep), the "roentgen equivalent mammal" (rem), and the gray.

Rep. Commonly used in radiation biology, this is defined as the dose delivered by 1 R to 1 g of water. Rep was initially defined as that dose of ionizing radiation which produces an energy absorption of 84 erg/g, and was later adjusted to 93 erg/g.

Rem. This is defined as the dose absorbed by a mammal when exposed to ionizing radiation which is biologically equivalent to the dose of 1 R of gamma radiation. The dose in rem is equal to the dose in rep multiplied by the "relative biological effectiveness".

Sievert. The dose of ionizing energy that produces the same biological effect on humans as a dose of 1 Gy from gamma rays or fast electrons. It replaces the older term, rem. For other forms of ionizing energy the relationship between the sievert and the gray is not one to one.

Gray. A unit of absorbed dose of ionizing energy. It replaces an older unit, the radiation absorbed dose, rad (the rad was defined as the absorption of 100 erg per gram of material). A radiation dose of 1 Gy involves the absorption of 1 J of energy of 1 kg of matter.

The conversion factors between radiation units are summarized in Table 4.

1.2 Interaction of Ionizing Radiation with Matter. Chemistry of Radiation

1.2.1 Interaction of Ionizing Radiation with Matter

The relevant interactions of gamma rays with food matter are primarily through two mechanisms: the photoelectric effect and Compton scattering (Fig. 3).

When a low-energy (below about 60 keV) gamma photon interacts with an atom some of the energy of the incident ray is used to remove an electron from an inner orbital, and the remainder is transferred to the electron in the form of kinetic energy. Thus, an ionization is produced. These photoejected electrons may carry considerable kinetic energy and are, therefore, capable of ionizing other molecules. There may be a whole string of secondary ionizations along its trajectory. The filling of the inner

Interaction of Ionizing Radiation with Matter

Table 4. Conversion factors between radiation units

$1\,\text{eV} = 1.6 \times 10^{-19}\,\text{J}$
$10^6\,\text{eV} = 1\,\text{MeV}$
$1\,\text{Gy} = 1\,\text{J/kg} = 10^7\,\text{erg/kg}$
$1\,\text{Gy} = 6.242 \times 10^{18}\,\text{eV/kg} = 0.24\,\text{gram-calorie/kg} = 100\,\text{rad}$
$1\,\text{kGy} = 100\,\text{krad}$
$10\,\text{kGy} = 1\,\text{Mrad}$
$1\,\text{rep} = 93\,\text{erg/g} = 0.93\,\text{rad}$
$1\,\text{rem} = 0.001\,\text{Sv}$

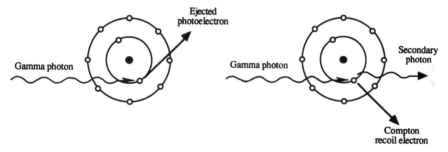

Fig. 3. Interaction of gamma rays with matter. *Left* The photoelectric effect; *right* the Compton effect

orbital vacancy from an outer orbital releases further energy, which may appear as characteristic X-ray fluorescence.

The Compton effect may occur when a gamma photon of somewhat higher energy (1 MeV) approaches a molecule. The photon gives up only part of its energy in ejecting an orbital electron and is deflected with a longer wavelength. The less energetic secondary gamma photon may then interact with another molecule, ejecting another electron. This process continues until all the energy has been transferred to electrons. The electrons, in turn, will each ionize other molecules and will produce paths of secondary ionizations.

Another possible process, such as production of an electron–positron pair, is important when the medium contains elements with a high mass number and for photons of energy greater than those employed in food treatment.

High-energy electromagnetic radiations lose a certain fraction of their energy in passing through each unit thickness of materials, and thus they never have a definite range. At any depth of penetration there is still radiation present, although the intensity becomes vanishingly small at great depths. The relationship for a narrow beam of monoenergetic gamma rays is:

$$\frac{I}{I_0} = e^{-\mu d},$$

where I/I_0 is the fraction of photons remaining in the beam after passage through an absorber of thickness d. The quantity μ is the total linear absorption coefficient and represents the absorption of photons by all of the processes described above. The absorption depends on the energy of the photon, on the density of the material, and on the atomic numbers of its components. The absorber thickness which is required to absorb half of the photons is given by the equation:

$$d_{50} = \frac{0.693}{\mu}$$

For example, the half-value thickness for 1 MeV gamma rays in lead is 0.90 cm and in air is about 25 m.

The study of radiation-induced chemical reactions of individual food components is an important source of information on the question of whether or not irradiated foods maintain their nutritive value, and on the possible formation of toxic and other deleterious substances. Ionizing radiation can induce chemical changes in a target material organic nature, either as a result of direct absorption of energy or indirectly, due to secondary reactions of the primary generated reactive species (Swallow 1977; Taub et al. 1979a).

Direct Effects. The principal characteristic of high-energy radiation is that, in contrast to any other segment of the electromagnetic spectrum, the energy per quantum of radiation is large enough to cleave molecules into electrically charged fragments and cause ionization in all materials. This property provides the distinction between ionizing radiation and its chemistry, and the next section of the electromagnetic spectrum, ultraviolet light and its chemical effects (photochemistry). The boundary is not sharp, but it can be taken as roughly equivalent to the energy of the most firmly bound outer electrons of 30 eV per molecule, corresponding to the wavelength of 40 nm in the vacuum ultraviolet. In photochemistry the energy absorption is quantized, i.e., usually a quantum of light of a narrow range of wavelengths undergoes a resonance interaction with a particular molecule of the mixture, to excite only one electron and perhaps break one specific bond. Because of this requirement for resonance, the interaction light-molecule is highly selective for a chemical configuration. Only ultraviolet light of very short wavelength carries enough energy to ionize certain molecules.

In radiation chemistry only part of the incident energy may be transferred at an interaction and the process is not selective (Spinks and Woods 1976). The energy is deposited randomly and nonselectively along the trajectory of the incident radiation causing ionizations (the amount of

Interaction of Ionizing Radiation with Matter

energy needed to eject an electron from a molecule ranges from 10 to 30 eV) until the energy of electrons is depleted to the level of vibrational excitations. In addition to ionization, the extra energy of the radiation is expended to form excited molecules. Thus, one incident photon may affect many thousands of molecules. Sometimes the excited molecules will dissociate, but the reactions are not necessarily the result of random cleavage of chemical bonds. There are preferred pathways which are largely influenced by molecular structure. In large molecules, e.g., proteins or carbohydrates, although the energy may be absorbed at one location it can be transferred, intra- or intermolecularly, to another, and at a "sensitive site" bond cleavage occurs. Provided that the medium viscosity is low and the transient species can diffuse readily, they spread out until a uniform distribution of primary radiolysis entities is attained. Since the radiation interacts nonselectively with the shell electrons, the energy deposited among the components of a mixture is nearly proportional to the number of electrons in the atoms of each component, or by a satisfactory approximation to weight fraction, and it is practically independent of the molecular configurations.

Indirect Effects. Once primary transients are formed by excitation or ionization, they may react with other molecules present in the mixture, depending upon relative reactivities (represented by the reaction rate constants) and concentrations. When an aqueous–organic system is irradiated, the nature of the change depends on the moisture content and, for moist foods, whether it is irradiated in a frozen or liquid state. If the water content is high, most primary reactive species are formed by the radiation-induced decomposition of water and the predominant indirect reaction of the organic molecule is the removal of a hydrogen atom by a hydroxyl radical, followed by various chemical conversions of the resulting secondary organic radical. At lower temperatures, the mobility of these primary radicals is limited and a recombination without net chemical change is more likely. Therefore, the possibility of an indirect action is decreased, resulting in a greater proportion of direct-type reactions of organic components.

In systems as complex as food, dose–rate effects involving primary radicals are not expected to be significant because pseudo-first-order reactions with the main components will predominate at almost all dose-rates used in practice.

The general framework of chemistry following the absorption of radiation is summarized in Table 5. When an electron is ejected from a molecule (A), the result is a positive ion (A^+) and a negative electron (Eq. 1). The electron cannot exist free for very long and is rapidly captured by another molecule (B) to yield a negative ion (B^-) (Eq. 2). The overall result of these events is the formation of an ion pair (Eq. 3). The ions last only a very short time (less than 10^{-10} s) and undergo one of the many possible

Table 5. The chemical framework of radiation

$A \rightarrow A^+ + e^-$	(1)
$e^- + B \rightarrow B^-$	(2)
$A + B \rightarrow A^+ + B^-$	(3)
$B^- \rightarrow E^- + F\cdot$	(4)
$A^+ + e^- \rightarrow A^*$	(5)
$A^* \rightarrow C\cdot + D\cdot$	(6)
$C\cdot + D\cdot \rightarrow A$	(7)
$C\cdot + F\cdot \rightarrow CF$	(8)
$D\cdot + O_2 \rightarrow DOO\cdot$	(9)
$D\cdot + R_1R_2C = CR_3R_4 \rightarrow R_1R_2DC - \dot{C}R_3R_4$	(10)
$D\cdot + RH \rightarrow DH + R\cdot$	(11)

reactions to form free radicals (Eq. 4). The free electron may be recaptured by the positive ion of the molecule from which it was ejected (Eq. 5). The reformed initial compound has a higher energy content than that associated with normal stability and may dissociate, usually to free radicals (Eq. 6). Free radicals are almost always intermediaries between ionization and final chemical products. They are extremely reactive as a result of an unpaired electron in one of their outer orbitals. Two free radicals can combine to reform original products, with no net effect on the system (Eq. 7), or can form new ones (Eq. 8). In this respect normal oxygen, which is naturally a stable biradical, is extremely reactive in combining with organic free radicals to form peroxyl radicals (Eq. 9). Radicals can also react with ordinary molecules by addition to unsaturated bonds (Eq. 10) or by hydrogen atom abstraction (Eq. 11).

The efficiency of a radiation-induced chemical transformation is expressed by the "G value" which represents the number of molecules of product formed per 100 eV of energy absorbed by the irradiated system. The amount of products formed from each component in a mixture would correspond to their G values in the pure system, normalized for the fraction of the total dose deposited in that component. This linear dependence of yields would be altered if: (1) energy transfer occurred from a component initially absorbing the energy to another component constituting a better energy trap; (2) there was a significant chemical interaction between components; or (3) if a minor component was not homogeneously distributed within a larger component. The latter occurrence, which is frequently experienced in food matrices, dictates a probability of reaction that is strongly dependent on local, rather than total, concentration.

Roughly, a dose of 10 Gy of gamma rays will produce a concentration change of approximately $1 \mu M/kg$, if the G value for the transformation is 1, and proportionally more if G is higher [G = 1 means that for 100 eV one molecule is changed. Therefore, for 10 Gy (6.242×10^{19} eV/kg) $6.242 \times 10^{17}/6.025 \times 10^{23}$, i.e., $1.036 \mu M/kg$ are transformed].

1.2.2 Radiation Chemistry

Water. In moist foods water is the macrocomponent most prone to direct radiolysis. The radiolysis of water yields several major decomposition products: hydrated electrons, hydrogen atoms, hydroxyl radicals, hydrogen, and hydrogen peroxide (Matheson and Dorfman 1965; Draganic and Draganic 1971; Hart 1972).

$$H_2O \rightarrow e^-_{aq}, H\cdot, OH\cdot, H_2, H_2O_2.$$

The G values for these transformations at $25\,°C$ are: $G(e^-_{aq}) = 2.65$, $G(H) = 0.6$, $G(OH) = 2.65$, $G(H_2) = 0.45$, and $G(H_2O_2) = 0.75$. Ice has a long dielectric relaxation time which promotes the return of radiation-ejected electrons to parent positive ions before hydration and, consequently, at $-5\,°C$ $G(e^-_{aq}) = 0.3$ only (in crystalline ice the formation of $OH\cdot$ is also inhibited, $G(OH) = 1$).

Among the products of water radiolysis, the hydroxyl radical is the most reactive and undiscriminating. Its rate of reaction with organic molecules is, for the great majority of chemistry, close to diffusion controlled. With unsaturated bonds the hydroxyl radical reacts by addition and with saturated molecules by hydrogen atom abstraction. In both cases, new organic free radicals are formed which sustain further reactions. The other products of water radiolysis are much more selective and possess a sensible lower reactivity.

Lipids. Irradiation of lipids results in nonoxidative (direct) and oxidative (indirect) changes.

Nonoxidative radiolytic changes of lipids are due to homolytic cleavages of interatomic bonds (Fig. 4).

Some bonds (a–e) are more susceptible to radiation and break easier than others (Nawar 1977). The resulting free radicals are terminated either by hydrogen atom abstraction from other molecules or by loss of a hydrogen atom and, to a lesser extent, by recombination with other radicals. The main stable products formed are H_2, CO_2, CO, alkanes, alkenes, and aldehydes.

The autoxidation of lipids is an indirect effect of radiation. This is a free radical chain process which can be initiated by any nonspecific free radical

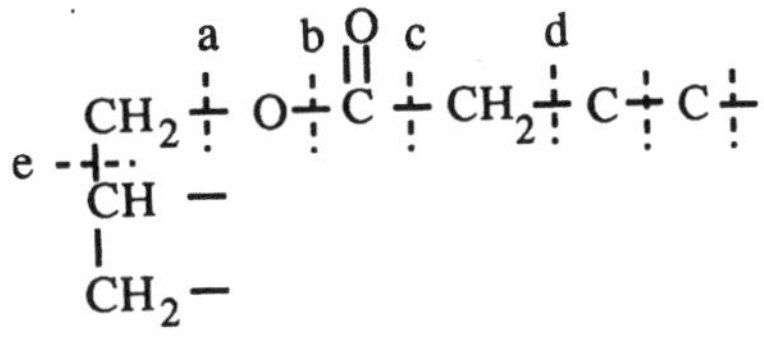

Fig. 4. Radiolytic cleavage sites on a molecule of fat

source, and ionizing radiation is no exception. After the initiation step, the further mechanism is always the same, yielding peroxides, alcohols, carbonyl compounds, hydroxy and keto acids, lactones, and polymers. The chemistry of lipid oxidation is discussed in Chapter 2. The autoxidation process can proceed over an extended period of time after irradiation. Unsaturated fatty acids are more easily oxidized than saturated acids. The usual parameters of fat, such as acid number, peroxide value, melting point, viscosity, density, and refractive index, are modified as a result of oxidation. The immediate effect of lipid degradation in foods is reflected in a downgrading of organoleptic quality. In this respect, the oxidation causes fatty foods to become rancid because of the formation of carbonyl compounds. After exposure to a dose of 20 kGy meat fat develops a distinctive, repugnant flavor, milk fat tastes chalky, and fish lipids smell unappetizing. Such chemical reactions may be minimized by irradiating food products when frozen, and by packing to exclude oxygen and light.

Proteins. Radiolytic modifications of proteins are the result of direct and indirect effects (Fig. 5).

The reactions of the water radical products, e^-_{aq} and the hydroxyl radical, with the simpler aliphatic amino acids in oxygen-free solution yield ammonia, keto acids, and fatty acids as the major products (Willix and Garrison 1967) (Fig. 6); these are illustrative of the chemistry involved.

The labile imino acid derivative produced in the disproportionation steps hydrolyzes spontaneously:

$$H_2O + NH^+_2 = C(R)COO^- \rightarrow NH^+_4 + RCOCOO^-.$$

The principal effect of irradiation on proteins is the splitting of large molecules into smaller units and therefore it is related to the structure of the particular protein: fibrous, globular, native, denatured, wet, or dry (Schaich 1980a). The hydrogen bonds of the secondary and tertiary struc-

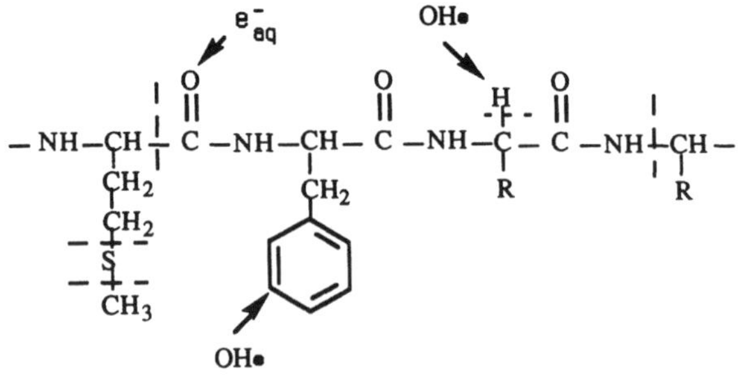

Fig. 5. Radiolytic cleavage of proteins

Radiation Chemistry

$$e_{aq}^- + NH_3^+CH(R)COO^- \rightarrow NH_3 + \overset{\bullet}{C}H(R)COO^-$$

$$\overset{\bullet}{O}H + NH_3^+CH(R)COO^- \rightarrow H_2O + NH_3^+\overset{\bullet}{C}(R)COO^-$$

$$\overset{\bullet}{C}H(R)COO^- + NH_3^+CH(R)COO^- \rightarrow CH_2(R)COO^- + NH_3^+\overset{\bullet}{C}(R)COO^-$$

$$\overset{\bullet}{C}H(R)COO^- + NH_3^+ \overset{\bullet}{C}(R)COO^- \rightarrow CH_2(R)COO^- + NH_2^+=C(R)COO^-$$

$$2 NH_3^+\overset{\bullet}{C}(R)COO^- \rightarrow NH_2^+=C(R)COO^- + NH_2^+CH(R)COO^-$$

Fig. 6. The reactions of e_{aq}^- and the hydroxyl radical with aliphatic amino acids

ture are weak bonds which can be broken in the vicinity of an ionization because the sudden introduction of a charge disrupts electrical dipoles.

Proteins in foods are generally well protected from ionizing radiation by other food components and at the doses employed in food irradiation the effect is minimal. Enzymes usually survive exposures of up to 10 kGy. The composition of amino acids scarcely changes as the result of irradiation and there is virtually no nutritional loss, though minor changes in the sulfhydryl groups of proteins have occasionally been noted (Taub et al. 1979b). Albeit in principle, the biological value of foods containing sulfur-rich proteins may be slightly lowered by irradiation; in practice, such deterioration leads primarily to an organoleptically unacceptable product, of which milk is a prime example.

Carbohydrates. On irradiation, hydrolysis and oxidative degradation of carbohydrate molecules may occur. Lower saccharides may be oxidized at the end of the molecule to form acids, and as a result of ring scission aldehydes may be formed. Large carbohydrate molecules are split into smaller units by cleavage of the glycosidic link, resulting in depolymerization. Thus, chains of pectic carbohydrates may be shortened with a loss of gelling power. However, the extrapolation of the results obtained with pure carbohydrates to real food stuffs requires caution because of inhibitory effects by other constituents in the food matrix.

Vitamins. The radiosensitivity of microconstituents, like vitamins, differs depending on whether they are in pure solution or in a food. The destruction of vitamins is, in most cases, indirect; free radicals of the solvent or oxidizing species, e.g., peroxyl radicals or carbonyl compounds, react with vitamins. Therefore, the percent of destruction is directly related to the content of water and oxygen.

Among the fat-soluble vitamins, vitamin E is the most radiosensitive and vitamin D the least. Vitamin A is also relatively sensitive to ionizing radiation since its activity is decreased by cis–trans isomerization. There are conflicting reports concerning the stability of vitamin K.

Thiamin is not only the most heat labile, but also the most radiolabile in the B group; pyridoxine is also fairly sensitive. Riboflavin, niacin, pantothenic acid, biotin, and folic acid are relatively stable to irradiation. Ascorbic acid is readily radioconverted to dehydroascorbic acid; as these compounds have similar vitamin C activity, this conversion is nutritionally insignificant. However, losses of vitamin C in irradiated foods have occasionally been reported.

Minerals. The minerals, as such, are unaffected by radiation. However, the mineral content of packed dry foods sterilized by ionizing radiation may be greater than that of heat-sterilized foods, since in the latter minerals may be lost in juices during cooking.

1.3 Applications of Ionizing Radiation in Food Technology

As early as 1921, food irradiation was used to kill the human parasite *Trichinella spiralis*. These parasitic nematodes live encysted in the striated muscle of hogs and during the digestion of infected meat the cysts with larvae are released in the stomach of the host, e.g., man, and cause trichinosis, a presently incurable disease. The symptoms of the disease, diarrhea, fever, muscle stiffness, and cysts, vary with the level of infection.

The advent of nuclear reactors in the 1940s made available large quantities of radioisotopes, such as cobalt-60 or cesium-137, at relatively low cost. At the same time, Van de Graaff generators and linear accelerators, which produced high-energy electron beams, became available. The positive pursuit of peaceful use of such materials and equipment sparked off research programs in many areas. Studies aimed at sterilizing food by irradiation began in the United States at the Massachusetts Institute of Technology in 1943 under the guidance of Prof. B.E. Proctor, and soon after the war more and more laboratories around the world followed these early research efforts. It was soon realized that sterilization of food requires doses of up to 50 kGy to eliminate heat resistant spores such as those of *Clostridium botulinum* and this level of irradiation produced unacceptable changes in the color and flavor of treated food. This, and the difficulty of finding an agreed protocol for testing the wholesomeness of irradiated foods, led to a decline of interest in radiation technology. In the late 1960s, however, the potential for using lower doses of radiation as a method for pasteurization, sprout inhibition in potatoes and onions, and destruction of insects in stored grains, began to be explored. The increasing public concern about the use of synthetic additives and the presence of chemical residues in food, as well as the prevention of food poisonings, have exalted the interest in low-dose irradiation of food. However, commercial applications have been scarce, although most of the relevant technical questions had been answered by the early 1970s. A possible formal reason might have been the insistence by the Food and Drug Administration (FDA) in the United States that ionizing radiation should be treated not as a food process of general applicability but as a food additive. This ruling required individual testing and approval of every item of food subjected to irradiation. The situation changed in 1976 when a joint committee of the Food and Agriculture Organization (FAO), International Atomic Energy Agency (IAEA) and World Health Organization (WHO) stated that irradiated potatoes, wheat, chicken, papaya, and strawberries are unconditionally safe for human consumption; the committee also gave their provisional approval for irradiated rice, fish, and onions. In 1986 the FDA amended its regulations to permit the use of ionizing radiation to inhibit the growth and maturation of fresh foods and to disinfect food of arthropod pests at doses not to exceed 1 kGy, and to disinfect dry or dehydrated aromatic vegetable substances, such as spices and herbs, of microorganisms at doses not to exceed 30 kGy (FDA 1986). Extensive research has been carried out during the past 30 years on the effects of ionizing radiation on foods and thousands of scientific reports have been published. The accumulated data indicate that ionizing radiation has some potential for applications to food production, but also has limitations of an economical, technological, and consumer acceptability nature.

The suggested applications in food technology are categorized according to the radiation dose employed; this generally varies with both the objective of the treatment and the type of food (Table 6).

The literature on food processing by radiation has been extensively reviewed (Elias and Cohen 1977, 1983; Urbain 1978b; Anon. 1983; Josephson and Peterson 1983; Overview 1989). In addition, a great deal of information has been collected in the specialized publications of the FAO of the United Nations and of the IAEA in Vienna.

Inhibition of Sprouting or Germination in Certain Crops (Onions, Potatoes, etc.). Potatoes and onions are among the most economically important vegetables grown and exported in many countries of the world. Therefore, major amounts of these food commodities are stored for long periods of time. Both vegetables, when mature and properly dried, usually maintain a good quality after long storage because of their low metabolic activity. However, prestorage practices during growing and harvest, as well as postharvest conditions, can influence the spoilage losses. Sprouting during storage is a major physiological cause of spoilage. There are several conventional options for sprout inhibition: postharvest application of chemical

Table 6. Applications of food irradiation in food processing

Low dose (less than 1 kGy)
    Inhibition of sprouting or germination in certain crops (onions, potatoes, etc.)
    Delay of senescence or control of ripening of some tropical foods
    Killing insects in cereal grains, fruit, cocoa beans, and other crops
Medium dose (1–10 kGy)
    Killing food-poisoning bacteria, particularly *Salmonella* and *Campylobacter*, in
        raw poultry, prawns and shellfish
    Killing parasites such as *Trichinella spiralis* and *Taenia saginata* in raw meat
    Extension of product life by reduction of microbial populations (by *ca.* 10^6) that
        spoil meat, fish, fruit and vegetables
Improvement of technological properties (increased juice yield from fruits,
    reduced cooking times for dehydrated vegetables, etc.)
    Sterilization of insects and parasites
    Sterilization of packaging materials
High dose (greater than 10 kGy)
    Sterilization of food
    Reduction of bacterial contamination of herbs and spices (by *ca.* 10^6)
    Reduction in the number of viruses (by *ca.* 10^6)
    Enzyme inactivation

inhibitors such as isopropyl-n-phenylcarbamate (IPC) or methyl ester of naphthalene acetic acid (MENA) on potatoes; preharvest spraying of onions with maleic hydrazide or ethepon (2-chloroethyl phosphoric acid); and low temperature storage.

When tuber, bulb, and root vegetables are treated with ionizing energy, morphological and histological changes in dormant buds induce necrosis at these growing points during subsequent storage. There is a great deal of information on sprout inhibition of potatoes (Thomas 1983) and onions (Thomas 1984) by ionizing radiation. Beneficial results are usually accomplished by doses limited to 0.05–0.15 kGy. Unlike the chemical sprout control, the inhibition achieved by irradiation is irreversible. A dependency of potato variety to required dose for sprout inhibition has been observed. Doses above 0.15 kGy may decrease significantly the wound-healing ability of potatoes (periderm formation) and makes them more susceptible to rot. Also, an increased sugar content, occasional incidence of black spots, and darkening after cooking have been reported. The data accumulated on onions appear to indicate a lesser variety–dose dependency than that observed with potatoes. In onions, a discoloration of the growth center, which is aggravated by higher doses, may occur during storage and cause internal rotting. It is recommended that irradiation should be performed as quickly after harvest as possible, since the treatment is most effective when applied before sprouting has been initiated. The biochemical mechanism of inhibition is explained by the radiation-induced modifications of hormone metabolism, as well as of synthesis of nucleic acids.

Applications of Ionizing Radiation in Food Technology 23

Sprouting of other root vegetables, such as garlic, shallots, yams, ginger, and turnip, can be similarly delayed by radiation.

Delay of Senescence and Ripening Control of some Plant Foods. Senescence is the phase of plant growth from full maturation to death, especially of the fruit and leaves, and is characterized by an accumulation of metabolic products, an increase in respiratory rate, and a loss in dry weight. A delay of senescence in fresh vegetable foods would enable an extension of product shelf-life and a reduction in spoilage losses. Its inhibition is another facet of possible changes induced by ionizing radiation in the postharvest metabolism of living foods. The extent of alteration of senescence by irradiation depends on the physiology of the plant food at the time of irradiation and on its postirradiation metabolism.

Climacteric fruits are those which ripen postharvest and, hence, undergo marked increases in rates of respiration and production of ethylene. The increase in consumption of oxygen and evolution of carbon dioxide is accompanied by all the symptoms of ripening: change in pigmentation from green to yellow or red, softening of the texture, development of aroma, increase in sweetness, and decrease in astringency. The duration and intensity of this process depends on the fruit maturity at harvest, and on storage conditions (e.g., temperature, composition of the atmosphere). If climacteric fruits are irradiated before the onset of the climacteric, the inhibitory effect may be very pronounced. Once the ripening process of a climacteric fruit has been initiated a normal radiation dose does not inhibit it any more. As a matter of fact, some climacteric stone fruits, such as peaches or nectarines, were even stimulated to ripen when irradiated in the preclimacteric state by doses of 1 kGy. A few tropical and subtropical climateric fruits that have been reported to undergo a delay in ripening after irradiation are bananas, mangos, payayas, guavas, and avocados. Delay in ripening of fruits is usually achieved with doses of 0.2–0.5 kGy. Some temperate-zone fruits, such as pear and apple, require higher doses (up to 1 kGy) for effective inhibition of ripening. Browning or scalding of the fruit skin, internal browning disorders in apples and avocados, and increased sensitivity of the fruit to chilling are possible postirradiation effects. The depolymerization of pectins, cellulose, hemicellulose, and starch in response to radiation results in softening which is undesirable for postharvest handling. There is a poor understanding of the biochemical mechanisms underlying the delay in senescence of climateric fruits by gamma irradiation. A multitude of hormonal and cellular changes are associated with the ripening process and it is difficult to establish an accurate relationship with the radiation treatment (Thomas 1985, 1986b).

Nonclimacteric fruits, such as citrus and grapes, reach desirable ripeness on the tree and do not undergo rapid metabolic changes after harvesting. The ripening of such fruits is stimulated by irradiation, probably due to the

24 Ionizing Radiation

induced generation of ethylene by radiation. Therefore, the treatment of nonclimacteric fruits with radiation, if considered applicable, is aimed only at minimizing the spoilage decay caused by fungal pathogens (vide infra).

Killing of Insects in Cereal Grains, Fruit, Cocoa Beans, and Other Crops. The use of ionizing radiation for control of insect infestation in grains and grain products has also attracted a lot of attention (Lorenz 1975; Tilton and Brower 1987). The growing concern about the health hazards associated with residues of pesticides in foods and the resistance developed by insects to chemicals are only two of the reasons for the widespread interest in disinfestation by irradiation. The final objective of insect control by gamma irradiation is the inactivation of all the species present. This can be achieved by preventing the emergence of the larva from the egg, preventing maturation during the pupa–adult stage, killing the larva, inhibiting reproduction, etc. Most insects are sterilized at doses of 0.05–0.75 kGy; some will survive 1 kGy, but their progeny are sterile. In general, eggs are the most sensitive to ionizing radiation, followed by larvae, then pupae. A dose of 0.5 kGy was found to inactivate all beetles and the immature stages of all moths, and surviving adult moths would not reproduce effectively after such an exposure. Therefore, this dose was recommended to control insects in bulk wheat, wheat flour, cornmeal, peanuts, nuts, dried fruits, rice, and leguminous seeds. This level of irradiation of grain offers also a side benefit: it produces measurable changes for the better in the baking qualities of the flour due to partial degradation of starch and proteins. Thus, the rheological properties, stability, and elasticity of the dough made from radiation-modified flour are improved. However, doses greater than 0.5 kGy impair the baking qualities and consumer acceptance.

The combination of gamma radiation with heat (conventional or infrared/microwave radiation), hypoxia, or chemicals is synergistic and enables the use of lower doses, down to 0.2 kGy. However, the major weakness of bulk cereal irradiation lies in its inability to protect against reinfestation after treatment.

In general, no specific microbiological problems arise during storage of dry foods such as cereal grains. The low moisture content prevents the organisms that survive the treatment with ionizing energy from multiplying and spoiling the food. If, however, malpractice of storage of dry cereal grain should occur, and cereal products are stored at a high relative humidity, this may allow growth of molds that survived the low levels of ionizing energy used for disinfestation. More unfortunate in such a case is the possible production of mycotoxins. Once formed, aflatoxin can be detoxified only by very high doses of ionizing energy. Thus, Temcharoen and Thilly (1982) found that only after treatment with 50–100 kGy of gamma rays did peanut meal lose its toxic and mutagenic properties attributed to the aflatoxin B_1.

Applications of Ionizing Radiation in Food Technology 25

Fruits and vegetables can also become infested with insects. While such infestations do not always spoil the food, as may happen with grain foods, there is justification for trade barriers in order to prevent the spread of insects. Irradiation at doses below 1 kGy is an effective insect disinfestation treatment against various species of fruit fly, orange worm, potato moth, spider mites, scale insects, and other insect species of quarantine significance in marketing fresh fruits and vegetables. Disinfection of the mango weevil, *Sternochetus mangiferae*, and of the oriental fruit fly, *Dacus dorsalis*, from Hawaiian papaya can also be achieved by less than 1 kGy radiation. Indeed, the irradiation of fresh fruits and vegetables as a quarantine control measure in international trade is one of the most likely uses of radiation treatment.

Killing Food-Poisoning Bacteria in Raw Poultry, Eggs, Red Meat, Prawns, and Shellfish. Moist human foods and related materials may contain a variety of pathogenic microorganisms which are relevant to public health. Some of these, such as *Salmonella*, *Campylobacter jejuni*, *Escherichia coli*, *Vibrio holerae*, *Vibrio parahaemolyticus*, and *Yersinia enterocolitica*, are well known. *Salmonella* spp., for example, not only contaminate raw foods of animal origin but are also present in animal feeds, such as fishmeal and byproducts, and these can easily serve as vehicles of contamination of human food, like from cattle feed to milk.

The purpose of radicidation is to destroy these pathogenic, nonspore-forming bacteria, short of complete sterilization of foods or animal feeds.

Microbial resistance to irradiation is expressed in terms of decimal reduction dose or "D-value". The semilogarithmic plot of dose–survival is usually a straight line after an initial shoulder or lag period, indicating a first-order death rate. From the straight-line segment, the decimal reduction dose for the specific microorganism can be calculated. This is the dose required to effect a 90% reduction (one log cycle) in the microbial population. The different microorganisms differ in susceptibility to ionizing energy and are characterized by different "D-values". There is also a variation in radiation sensitivity with the strain and with the properties of the host medium, like its oxygen and moisture content, whether or not it is in a frozen state, etc.

The elimination of nonsporing pathogens from moist foods, such as fresh meat, poultry, fish, and seafood, requires doses of 3–10 kGy (El-Zawahry and Rowley 1979; Lambert and Maxcy 1984; Tarkowski et al. 1984; Palumbo et al. 1986) and there are some concerns for the consequences on product quality of applying such doses.

Killing Pathogenic Parasites in Raw Meat. The attempt to kill worms capable of causing diseases was one of the primary intended applications for irradiation of foods (Engel et al. 1988). Although low doses (0.2–0.3 kGy) inhibit reproduction and maturation of *Trichinella spiralis*, the efficiency of the treatment was doubted since it may not prevent the initial phase of the disease associated with the release of the ingested organisms in the intestines. On the other hand, the dose of irradiation needed to cause the death of *Trichinella spiralis* in pork, *Cysticercus bovis*, *Cysticercus cellulosae*, and *Cysticercus pisiformis* in beef, or *Anisakis* larvae in salted herring, requires from 3 to more than 6 kGy. Such relatively high doses are suspected of inducing objectionable sensory changes in fatty foods.

Extension of Product Life by Reduction of Microbial Populations that Spoil Meat, Fish, Fruit, and Vegetables. The objective of radurization is to extend the shelf-life of a food product by reducing the microbial load originally present by *ca.* 10^6. The irradiation dose for radurization should be optimized for minimal undesired changes of flavor, which limit the upper dose, and the desired product life, which limits the lower dose. Since the products are not sterilized after treatment, to attain the desired product life the radiation is complemented with refrigeration. Other spoilage factors, like chemical changes due to lipid oxidation or reactions induced by endogenous enzymes, are unaffected by radurization. Radurization is considered to be particularly useful for fish, seafood, poultry, and eventually strawberries.

Fish and Seafood. The perishability of fish and seafood is primarily due to microbial spoilage and, indeed, radurization with doses below 5 kGy is effective in significantly extending the shelf-life of these products, when refrigerated. Not surprisingly, the best results are obtained with irradiation of freshly caught fish. Thus, a dose of 0.5 kGy applied immediately was as effective as a five-times larger dose applied 9 days later.

The composition of microbial populations is altered by the treatment, since not the entire microflora is equally sensitive to radiation. *Pseudomonas* spp. are particularly sensitive to irradiation and their elimination contributes significantly to the sensory acceptability of irradiated fish. On the other hand, the selective reduction in the microflora of fish and seafood may favor the competitive development of *Clostridium botulinum*, as the radurization does not inactivate the spores of this organism. As long as foods are refrigerated below 10 °C there is no microbiological safety problem from *Clostridium botulinum* types A and B. Type E, however, can grow and produce toxin even under refrigerated conditions of 3 °C or above. The extension of the storage period due to radurization potentially enhances the opportunity for toxin production without the usual warning signs of spoilage. The extent of such a hazard depends primarily upon the initial level of contamination, the dose level of radiation used, and the temperature of storage. In products eaten raw, such as smoked herring fillets

Applications of Ionizing Radiation in Food Technology 27

or smoked salmon, the potential hazard is more real. However if good practice is followed in the processing line, along the distribution chain until consumption, and, in particular, if the temperature of the product is never allowed to exceed 3 °C, there should be no danger of botulism poisoning. Synergistic effects of other physical or chemical agents which are additive to irradiation, like heating or salting, might minimize the botulism hazard.

Fatty fish, such as mackerel, herring, salmon, and trout, are not suited to radurization because of their susceptibility to lipid oxidation and color changes.

Poultry and Other Fresh Meat. Due to the ease of cross infection the surface of meats is contaminated during cutting, evisceration, and preparation. Radiation could delay microbial spoilage, in particular by destroying the very radiation-sensitive Gram-negative rods, *Pseudomonas*, *Achromobacter*, and *Flavobacterium*. Subsequently, Gram-positive organisms, either anaerobes or facultative anaerobes, like lactobacilli, outgrow the radiation-sensitive microorganisms. Other deterioration changes of fresh meat, like oxidation of fat, cannot be controlled by irradiation. In fact, observations that radiation itself promoted fat oxidation and caused discoloration have been reported.

Fruits and Vegetables. The radurization of fresh fruits and vegetables has been studied extensively (Thomas 1986a,b). Fruits and vegetables are protected against microbiological spoilage by epidermis. However, accidental wounds incurred during harvesting and handling facilitate contamination and subsequent infection. Filamentous fungi are the primary cause of fruit spoilage; the bacterium *Erwinia carotovora* is the primary cause of soft rot of vegetables.

It seems, however, that radurization is, in general, unsuitable for controlling diseases of fruits and vegetables. The organisms that cause rot and deterioration can only be controlled by radiation dose levels that are near or above the tolerance level for the tissue of the fruit. Such doses induce downgrading changes in quality, like softening, texture loss, flavor changes, etc; citrus fruits, for example, are particularly susceptible to skin pitting. In order to decrease the radiation doses below the threshold of damage to fruit, combination treatments with dipping in hot water or application of fungicides have been suggested. The cumulative action of different factors may decrease the requirements for each of them taken separately.

There are only a few cases, however, such as strawberries and fresh figs, when development of spoilage organisms can be delayed to advantage solely by irradiation treatment. For example, fresh strawberries, when stored, often develop a white-gray, cotton-like mold growth caused by *Botrytis cinerea* that softens the tissues. It has been demonstrated that

28 Ionizing Radiation

application of a radiation dose of 2 kGy prevents small lesions and delays the appearance of larger lesions. The shelf-life of fresh strawberries may be extended by 5–8 days when stored at 5 °C or lower. In view of the very short market-life of untreated berries this extension is of significant commercial value.

Milk and Dairy Products. The key to a long shelf-life for milk and fresh dairy products is the prevention of microbial spoilage. In this category of foods, which is mainly aimed at the market of babies and children, and strives to preserve a healthy and natural image, the use of synthetic preservatives is very restricted. Nevertheless, the possible use of radiation as a physical alternative to chemical additives is unlikely. The dairy products are very sensitive to radiation; a disagreeable flavor is produced even at doses as low as 0.5 kGy, which makes any potential application improbable. It is noted that the minimum radiation dose for sterilization, 12D, for *B. cereus*, preinoculated into cheese and ice cream and gamma irradiated at −78 °C, was 43–50 kGy (Hashisaka et al. 1990).

Improvement of Technological Properties. The chemical modifications of some food constituents by radiation may be beneficial for further processing. Although it is not likely that this treatment will gain acceptance for such a purpose, since there always alternative practices available, a few potential applications are surveyed here.

The molecular degradation of proteins and carbohydrates can make meat and vegetables less tough in texture; radappertized meat becomes tender. Dehydrated vegetables rehydrate more readily after irradiation. Doses of up to 4 kGy increase the drying rate of blanched prunes, and the juice yield from grapes rises from 2 to 28% in proportion to an increasing radiation dose from 0.5 to 16 kGy (Sowden 1981).

The degradation of natural carbohydrates in legumes by irradiation is another potential application. Dry beans constitute an important source of inexpensive food proteins. However, their consumption is limited due to the inconvenience created by the "hard-to-cook" phenomenon, such as the need for long soaking and cooking times, and also the association with flatulence. Efforts to improve the cooking quality of dry beans have included various pretreatments, including gamma radiation (Rao and Vakil 1983). Irradiation of hydrated seeds with 2.5 kGy caused 50% reduction of both stachyose and raffinose, the two most gas-forming sugars (these non-reducing sugars cannot be completely metabolized due to the absence of α-1,6-galactosidase activity in mammalian intestinal mucosa), and 20% loss in the total content of oligosaccharides. Quantitative analysis of the breakdown products suggested that radiation treatment stimulated the glycosidic cleavage of higher sugar molecules to simpler, more easily digestible sugars.

Applications of Ionizing Radiation in Food Technology 29

Sterilization of Insects. There are two practical applications of radiation to the problem of insect control in agriculture. One is by exposing insects to annihilating doses of radiation, as in the case of stored-product insects (vide supra), and the other is by release of sterile insects into the environment. The latter procedure consists essentially of releasing large numbers of laboratory-bred irradiated males into the natural population, so that when sterile males mate with sexually normal wild females, the latter fail to produce offspring. With appropriate competition by the sterile males, most of the females will lay sterile eggs. By repeating this process for a few generations the pest population is virtually eliminated. Sterility is induced in the male by exposing pupae or young adults to a source of gamma radiation such that the sterilizing dose does not adversely affect their mating and inseminating ability.

Sterilization of Food. The achievement of indefinite shelf-life stability of a food product without refrigeration requires the total destruction of spoilage organisms and pathogens, as well as the inactivation of endogenous enzymes which could cause undesirable chemical changes. The dose requirement for radappertization is determined by the most radiation-resistant food microorganism, which is not necessarily also the most heat-resistant. Indeed, the most resistant organism of concern to radiation is *C. botulinum*. The radiation resistance of its spores varies with both the bacterial strain and the type of infected food. Some other microorganisms whose radiation resistance is greater, such as *Acinetobacter* and *Moraxella*, are readily killed by a preirradiation thermal treatment at 65–75 °C, and do not constitute a factor in the production of radappertized foods. This preirradiation heating treatment is anyway needed to inactivate autolytic enzymes present in certain meats and fish, since their deactivation is much more easily achieved by thermal treatment than by radiation (Welch and Maxcy 1975).

The minimum radiation dose for sterilization is defined in accordance with the 12D concept of microbiological safety which is a reduction in the number of viable spores by a factor of 10^{12} (Thayer et al. 1986). The dose values required to achieve this reduction depend on the food and its temperature during irradiation. The irradiation temperature optimized for minimal formation of unpalatable flavors due to protein decomposition sems to be -30 °C. The dose that is usually employed for radappertization is about 45 kGy. The foods which contain additives like sodium nitrite $NaNO_2$, sodium chloride NaCl or calcium chloride $CaCl_2$ need slightly lower doses than foods without these ingredients. These additives, as well as some spices, mustard oil, and nutmeg, reduce the radiation resistance of *C. botulinum* in ground beef.

A sterilization treatment, and irradiation is no exception, must be combined with suitable packaging to prevent recontamination. In this respect,

30 Ionizing Radiation

the experience accumulated with thermal methods of sterilization (canning, ultra-high temperature products) has been most useful. Metal and multi-layered plastic films have also been found to be satisfactory for radappertized products. The use of glass is aesthetically objectionable since it is discolored by radiation (electrons trapped in the glass matrix render it a brownish color).

Among the products that have been developed are radappertized meat products of beef, pork, chicken, fish, and seafoods. Nevertheless, the high doses of radiation required for radiosterilization result in some undesired side effects such as formation of unpleasant odor and flavor and changes in texture and color. On the other hand, the radiation-induced changes in protein molecules have a tendering effect and radappertized meats may have superior texture characteristics and a restrained moisture release, as compared to thermally sterilized meats. However, the final test of consumer acceptability can be made only when such products become available to consumers on a basis that allows open competition in the market place.

Rather specialized applications of radappertization include the preparation of sterile diets for hospital patients with a low immune response, usually after organ transplant, and who are housed in a germ-free, protected environment and must be maintained on a low microbial diet. The radappertization creates the possibility of a more diverse diet, which contributes to better healing. Likewise, sterile foods are needed by astronauts and pathogen-free laboratory animals.

Reduction of Bacterial Contamination of Herbs and Spices. Many spices are highly contaminated by bacteria and fungi originating from the plants or soil. The microbial load of spices increases during harvesting and processing. The majority of the microbial flora of spices consists of aerobic spore-forming bacteria, usually bacilli and molds. Sometimes coliforms and streptococci can be found, but clostridia, lactobacilli, micrococci, staphylococci, and yeasts are rare. The contamination is particularly heavy in spices, such as pepper, paprika, thyme, marjoram, mustard power, and mixed spices such as curry powder, and relatively low in cloves and cinnamon. Aerobic microorganisms plate counts of 10^6 bacteria and 10^4 molds are typical.

The objective of decontamination of herbs and spices is not the preservation of the product but rather the reduction of the indigenous population of microorganisms which could contaminate and spoil other foods in which the spices are incorporated. The two available control methods for eradicating microorganisms are chemical fumigation and irradiation. Thermal decontamination of spices is not a viable alternative because the heat drives off or destroys the desirable volatiles which characterize spices. The agent most widely used for killing insect pests in seasonings and reducing the viable microbial cell count of spices is ethylene oxide. To a lesser extent other fumigants, like propylene oxide and methyl bromide, are used too. Such chemical treatment has been criticized for causing toxicological problems.

The irradiation technique is particularly attractive for the treatment of spices since the common heat-resistant bacteria found in spices are sensitive to radiation and a dose of 4–7.5 kGy is generally considered to be sufficient to bring the microbial load to an acceptable level of 10^4 counts/g. This would not increase the bacterial count load of the food to which it is added. Vajdi and Pereira (1973) demonstrated in comparative studies that gamma radiation was not only more effective than the common fumigation agent, ethylene oxide, in reducing the bacterial population, but also preserved better the quality of spices. Thus, the oil content and the color of certain spices was insignificantly affected by radiation, and much less when compared to chemical treatment. The high doses of radiation needed to disinfect dry or dehydrated aromatic vegetables can modify the chemical structure of carbohydrate constituents, resulting in a shorter rehydration time for products such as onion flakes, beans, and peas. Radiation treatment of spices, unlike fumigation, can be run continually and can be applied to prepacked materials. Because of their remarkable stability and high value per unit volume, spices could be irradiated economically at an off-site facility.

Reduction in the Number of Viruses. Because viruses are resistant to radiation doses sufficient to eliminate them may be much higher than those tolerated by the quality requirements of the food products to be treated. Fortunately, viruses are sensitive to heat and can be eliminated from most foods by a heat treatment, combined eventually with irradiation.

There has been a mild interest in the use of radiation as a quarantine-control measure for the foot-and-mouth virus. This virus causes disease in meat animals and since at present certain areas of the world are free of this virus, embargoes on fresh meat from areas suspected of contamination are exercised to prevent its spread. Since the inactivation doses needed are high (40 kGy) and should be applied to fresh meat, it is virtually impossible to avoid production of unpalatable flavors.

Enzyme Inactivation. The enzymes are not very much affected by doses of ionizing radiation, which would be inoffensive to the wholesomness of a food product. Therefore, even in radappertized products, the deactivation of endogenous enzymes is more easily achieved by a preliminary thermal treatment.

1.4 Radiation Dosimetry

M. Faraggi[1]

In the technical specifications for the irradiation of various foods a dose range is recommended for each type of treatment. Dosage has to be expressed in terms of a range not only because it is very difficult to have a completely uniform spatial dose distribution, but also because an optimum amout of radiation cannot be fixed. The range encompasses all unavoidable variations in parameters of the same foodstuff for achieving a desired effect. The specification of a range implies that no part of the food shall receive less than the minimum dose, since very drastic safety aspects might be involved and, on the other hand, more than the maximum dose might adversely affect the quality. Therefore, careful dosimetry is a necessity in the irradiation of foods for human consumption. Satisfactory methods of dosimetry exist and have been used for years in commercial radiation technology (Whyte 1959; Holm and Berry 1970; Anon. 1977; McLaughlin 1982; Kase et al. 1987; McLaughlin et al. 1989).

Techniques for measuring radiation doses can be divided into two classes: physical and chemical methods.

1.4.1 Physical Dosimetric Methods

A dose of ionizing radiation may be monitored by three physical methods: calorimetry, ionization or scintillation.

Calorimetric Dosimetry. This method is based on the direct determination of the temperature increase of the target material, as the result of exposure to a certain amount of energy delivered by radiation. Most of the calorimetric methods (adiabatic, isothermic, or stationary) are suitable for this kind of dosimetry. Since a temperature variation of $10^{-5}\,°C$ can be measured, the sensitivity of the calorimetry is in the order of $0.1\,Gy$ ($1\,Gy$ produces a temperature rise of $2.39 \times 10^{-4}\,°C$) and dose rates ranging from $1\,Gy/h$ to $10^4\,Gy/s$ can be determined. An absorbed dose is converted to heat if none, or only minor chemical transformations (which are known and can be corrected for), are initiated by that dose. For example, with water, the most common target material, a correction of about 1% of the dose, due to the formation of H_2 and H_2O_2, should be made.

---

[1] Department of Chemistry Nuclear Research Center-Negev, P.O. Box 9010, Beer-Sheva 84190, Israel

Ionization Dosimetry. The measurements of gas ionization by radiation have been used for dosimetry since the discovery of X-rays and radioactivity. These methods determine the number of ion pairs created in air by the incident radiation (2.1×10^5 ion pairs in $1\,m^3$ correspond to $8.3 \times 10^{-3}\,Gy$). Ionization is measured in a gas-filled (air) ionization chamber with two parallel electrodes. The ions created are attracted to the electrodes by an electric field and the resulting current is measured. However, all the ions created by the incident radiation in a defined volume, including those from secondary electrons, are to be measured. This requires that the dimensions of the parallel electrodes should be at least double the length of the electron trajectory in air. This is feasible with soft X-rays and electrons but difficult to achieve for high-energy radiations.

Alternatively, the ionization measurement can be performed in a thimble gas-filled ionization dosimeter. In this device, a thimble air condenser whose walls are made from thin material with a stopping power similar to air, replaces the gas-filled (air) ionization chamble. "Air-equivalent" materials with similar electron concentration (mean atomic number), such as organic polymers (e.g., bakelite), do not modify the electronic equilibrium of the irradiated medium. Before use, the condenser is charged to an appropriate potential. When exposed to radiation the charges are partially neutralized. The number of ions produced by the ionizing radiation is measured by the potential drop and is equivalent to the radiation dose. This instrument offers a routine and sensitive means of measuring exposures to X-rays and gamma radiations, and also to linear accelerator beams provided that the photon energy is not too high.

Scintillation Methods. These are based on the property of certain materials to convert an incident photon of ionizing radiation to light. The light emission is proportional to the incident energy. By measuring this light output with a light meter, the irradiation dose may be calculated.

1.4.2 Chemical Dosimetric Methods

The physical dosimeters based on calorimetry or ionization are most frequently used for initial calibration of chemical dosimeters which are subsequently used for routine measurements. These are chemical reactions of a known radiochemical yield.

Ideally, a chemical dosimeter should have the following characteristics:

1. A good reproducibility within 1–5%. This results from a small influence of uncontrollable or difficult-to-control parameters during irradiation (sensitivity to minute amounts of impurities, chemical instability, or small dependence on temperature, light or air).

2. A proportional response over a wide range of absorbed doses (from a fraction of a gray to 10^6 Gy). This proportionality may be affected by reductions in reagent concentrations (which in some systems could be dissolved oxygen), changes in solution pH, accumulation of radiolytic products, etc.
3. Independence of parameters other than dose, such as dose rate (from a few Gy/min to 10^{11} Gy/s) and energy.

Solid-State Chemical Dosimeters. In commercial applications of radiation convenience is of prime importance and in this respect solid-state chemical dosimeters are most attractive. Radiation effects on solids, such as positive holes created by the departure of electrons trapped elsewhere in a solid matrix (F centers) and chemical reactions in solid state, have been employed for dosimetry. These processes are visualized as glass coloration, opacity of plastic materials, fogging of photographic films, and thermoluminescence. Solid anthracene is representative of the organic compounds whose luminescence has been used to monitor radiation. This compound fluoresces at 440 nm and upon radiation it degrades to nonluminescent compounds. Doses between 5×10^3 and 5×10^5 Gy and dose rates up to 40 Gy/s have been measured in this way. Other organic compounds, such as polymethylmethacrylate and polystyrene, are representative of dosimeters using near ultraviolet and visible spectrophotometry. In polymethylmethacrylate the formation of colored centers by doses from 10^3 to 10^5 Gy is measured as an absorption loss in the optical range of 260–345 nm. The radiation induced trans–cis isomerization of trans-stilbene in doped polystyrene was followed by the optical absorption of the trans-stilbene at 324 nm. An alternative system consists of films, papers, or solutions containing triphenylmethane radiochromic leuko dyes, such as hexahydroxyl ethyl pararosanile cyanide, which become colored on irradiation in the range 10–100 kGy. Other solid-state dosimeters are glasses (cobalt or the silver doped Li-Al phosphate) and p-on-n solar cells.

In addition to solid dosimeters a variety of chemical dosimeters are based on radiation effects in solution. In these systems the absorbed dose E (Gy) is calculated from:

$$E = 9.648 \times 10^6 \frac{\Delta M}{\rho G},$$

where ΔM is the concentration of the transformed species in mole/l and ρ is the density of dosimeter in g/cm^3.

Commonly, absorption spectroscopy is usually used to determine the concentration of the transformed species.

Among the liquid dosimeters the Fricke dosimeter and the Ce^{IV}/Ce^{III} system are the most common.

Chemical Dosimetric Methods

The Fricke Dosimeter. The reaction involved in the Fricke dosimeter is the radiation-induced oxidation of a ferrous ion to a ferric ion in the presence of oxygen in an acid solution of ferrous sulfate (or ferrous ammonium sulfate). The standard Fricke dosimeter is an air-saturated ($[O_2] = 2.5 \times 10^{-4} M$) aqueous solution which contains $1 mM$ ferrous ions ($FeSO_4$ or $(NH_4)_2Fe(SO_4)_2$), $0.4 M$ H_2SO_4 and $1 mM$ NaCl.

The concentration of the ferric ions formed after irradiation is determined by spectrophotometry by comparing the optical density of the irradiated and nonirradiated solutions at $\lambda = 304\,nm$. At this wavelength and at $25\,°C$, the molar extinction coefficient (ε) of Fe^{3+} is $2205 \pm 3\,M^{-1}cm^{-1}$ and that of Fe^{2+} is negligible. The $G(Fe^{3+})$ value for the Fricke dosimeter depends on the nature and energy of the radiation. However, for photons (X-rays and gamma rays) between 0.6 and 35 MeV and for electrons from 1 to 30 MeV (diffuse radiation) the value of 15.5 ± 0.2 is acceptable.

Filling in the equation with the ε value for Fe^{3+}, an optical path of $1\,cm$, and the density of the Fricke solution ($\rho = 1.024\,g/cm^3$), the absorbed dose is:

$$E = 276\,\Delta(OD)\ Gy$$

The upper limit of absorbed dose that can be measured with the aerated Fricke dosimeter is ca. $500\,Gy$. With this dose both Fe^{2+} and the dissolved oxygen are almost consumed. This limit can be increased to about $2000\,Gy$ by increasing the Fe^{2+} concentration to $20\,mM$ and saturating the solution with oxygen ($[O_2] = 1.2 \times 10^{-3} M$). The lower limit depends on the accuracy of the optical density measurement. Assuming a $\Delta(OD)$ of 0.1, the lower limit is about $30\,Gy$ with a $1\,cm$ cell and $3\,Gy$ with a $10\,cm$ cell. The dosimeter is also influenced by high dose rates. The upper limit for the normal Fricke dosimeter is $2 \times 10^6\,Gy/s$, and for oxygen-saturated solutions the dose rate can be increased to $10^8\,Gy/s$.

The Ce[IV]/Ce[III] Dosimeter. The "cerium" dosimeter is based on the radiation-induced reduction of ceric ions (Ce^{+4}) to cerous ions (Ce^{+3}) in acid solution. The G value of 2.4 for this transformation suggests a mechanism where the Ce^{+4} is reduced to Ce^{+3} by hydrogen atoms and hydrogen peroxide while hydroxyl radicals oxidize Ce^{+3} to Ce^{+4}. Therefore, the mechanism is independent of the presence or absence of oxygen and the upper limit for the radiation dose of $10^6\,Gy$ is confined by the solubility and consumption of ceric salts. The lower limit is about $50\,Gy$.

The reaction is monitored by measuring the decrease of Ce^{+4} absorption at $320\,nm$ ($\varepsilon = 5610\,M^{-1}cm^{-1}$) and it is practically independent of temperature. The fact that Ce^{+3} concentration, and therefore radiation doses, is measured by difference imposes an initial high Ce^{+4} concentration. It is

36   Ionizing Radiation

recommended that for doses of $100\,Gy$ a Ce^{+4} concentration of $0.1\,mM$ should be used. The concentration should be gradually increased for higher doses (for $10^6\,Gy$ the Ce^{+4} concentration should be $0.4\,M$).

Other chemical systems have been proposed and used as dosimeters, but none of them has been adopted as widely as the Fricke or cerium dosimeters. The reasons for searching for new systems are: the limited range of the Fricke dosimeter ($40–400\,Gy$); the great sensitivity to impurities and light of the Ce^{+4} dosimeter; and the fact that both are highly corrosive.

The dose range of the Fricke dosimeter was extended to $10^5\,Gy$ by adding $10\,mM$ $CuSO_4$, which reduces the G value from 15.5 to 0.65 and prevents oxygen consumption. Another system that was recently developed for industrial applications ($10–40\,kGy$ range) is the acidic dichromate dosimeter where the loss of $(Cr_2O_7)^{-2}$ (G = 0.4) is determined spectophotometrically at $440\,nm$. The oxalic acid dosimeter ($25–600\,mM$ in aerated aqueous solution) is based on the loss of oxalic acid as measured spectrophotometrically or by titration, and covers the range $10\,kGy–2MGy$. Organic dyes have been suggested as dosimeters and the bleaching of methylene blue solutions was one of the first used for many years. Finally, radiation-induced chain reactions were suggested as low-dose dosimeters. One example is the release of hydrochloric acid from organohalogen compounds such as chloroform. Since the G value is high (*ca.* 30) and the hydrochloric acid is easily determined with great precision, doses as low as mGy could be determined. However, because the chain reactions are very sensitive to impurities and changes of temperature, these chemical dosimeters are not considered very reliable.

1.5 Safety and Wholesomeness of Irradiated Foods

There are two different safety elements in the technology of food irradiation: the potential danger of operating the irradiation facility and the possible hazards presented by the irradiated food.

The exposure to ionizing radiation, which is possible only by a faulty operation of the radiation plant, presents a very serious health hazard. The ionizing radiation cannot be detected by any of the human senses, unless its intensity becomes very great and then it is sensed as heat. The direct and indirect impact of large doses of radiation upon living organisms results in a wide range of effects (Table 7).

At the molecular level the ionization events are held to be the principal molecular cause of biological damage, which includes chemical modifications of proteins, enzymes, nucleic acids, and components of cell membranes. These modifications are enhanced by the omnipresence of oxygen which reacts at the level of the initial chemical lesion and leads to the

Safety and Wholesomeness of Irradiated Foods

Table 7. Radiobiological damage in mammals

Level of damage	Radiation effect
Molecular level	Modification of biomacromolecules (proteins, enzymes, nucleic acids) and interference with metabolic pathways
Subcellular	Damage to cell membranes, nucleus, chromosomes, mitochondria and lysosomes
Cellular	Inhibition of cell division, cell death, and mutations
Tissue, organ	Disruption of central nervous system, bone marrow, and malignancy
Whole animal	Diseases and death
Populations of animals	Changes in genetic characteristics due to gene and chromosomal mutations in individual members of the species

appearance of more damaging free radical peroxy species. The primary chemical lesions are amplified to temporary or permanent disruptions of biological activities, mutations, diseases and eventually death. A dose of 5 Gy can kill a human being (but a simple virus can survive a dose of up to 200 kGy!). In view of these very serious risks the radiation facilities are equipped with safety provisions which, when properly operated, are virtually accident-proof. The machinery includes safety interlocking systems which are designed in such a way that failure of any electronic or mechanical components will not result in any accidental exposure of personnel to radiation.

Studies to prove the safety of foods treated with ionizing radiation involved the following aspects: induced radioactivity, nutritional adequacy, and toxicological and microbiological safety.

Avoidance of induced radioactivity is a basic requirement of the food irradiation process. Stable nuclei can be made unstable, or artificially radioactive, by the introduction of energetic particles such as electrons, neutrons, protons, etc. High-energy gamma rays are also capable of producing excited atoms which disintegrate by radioactive decay. To prevent the inducement of nuclear change in the constituent elements of the food, thus causing the food to became radioactive, it is necessary to limit the energy level of the radiation employed. This is accomplished by setting energy limits of 5 MeV for gamma and X-rays and 10 MeV for electrons. The gamma radiations emitted by nuclear disintegration of ^{137}Cs or ^{60}Co are intrinsically limited to 0.66 and 1.33 MeV levels, at most.

The artificial irradiation can contribute to radioactivity in food much less than do natural isotopes. Food is naturally radioactive because it contains small quantities of long-lived radionuclides, including ^{14}C, ^{40}K, ^{87}Rb, and ^{204}Pb. The natural radioactivity in meat, for example, amounts to about

100 Bq; a dose of radiation that would completely sterilize meat by killing off all bacteria could induce 10 Bq of radiactivity.

Three nuclei that occur regularly, but in very small quantities, in food can undergo nuclear modifications (neutron ejection) by radiation energies below 5 MeV: deuterium (^{2}H), a rare stable isotope of natural hydrogen (0.02% natural occurence); ^{17}O, a rare isotope of oxygen (0.04% natural occurence) and ^{13}C (1.1% of natural carbon ^{12}C). In each instance the nucleus which remains is not radioactive. The common elements, hydrogen, ^{16}O and ^{12}C, which predominate in living matter including food, require more energetic levels of radiation to become radioactive. The radioactive isotopes of nitrogen, oxygen, and carbon which can appear above the allowed levels under the influence of high-energy radiation are short lived, their half-lives being between 2 to 21 min.

Over the past years many wholesomeness tests have been carried out to prove that irradiation products are safe for consumption. Experiments on the response of experimental animals and humans have been conducted to investigate the nutritional quality of food treated with ionizing energy under conditions that could be used commercially. Tests to determine the utilization of the nutrients and clinical tests of the subjects disclosed no unfavorable effects of foods processed with ionizing energy relative to comparable foods processed by conventional means. The methods used for toxicological testing involved the incorporation of the foodstuff under test into the diet of laboratory animals, which were kept under observation over a long period of time. Factors like rate of growth, behavior, incidence of disease (including cancer), changes in hematological and biochemical parameters, chromosomal abnormalities, genetic changes, reproductive performance, birth defects in offspring, and effects on mortality were comparatively monitored with groups of control animals and were taken into account in assessing the effects of irradiated food. In none of the studies were any abnormalities observed which could be attributable to the irradiation treatment received by the food administered to the laboratory animals. Individual radiolytic products in treated food are normally present at very low levels and many, or all, of them are the same as those produced by cooking or other forms of processing. So far no toxic "unique radiolytic products" have been found.

Research on the influence of radiation on chemical constituents of food was also carried out on the five main components necessary in an adequate diet, namely fats, carbohydrates, proteins, vitamins, and water. Changes in chemical composition and nutritional value were monitored, and examination of the possible formation of toxic degradation products was conducted. Although ionizing energy may cause changes in the physical and chemical properties of carbohydrate and protein components, these changes are not nutritionally significant. No radiation-specific impairment of nutritional values attributable to the commercially recommended doses (below 10 kGy) could be established. Some vitamins, like tocopherols, are radiation sensitive but they are also sensitive to heat to a similar extent. Thus, the alterations in irradiated foods are partly of the same kind as in conventional processing methods, including changes in composition, vitamins color, and are partly specific to the irradiation process. The deterioration effects attributed to radiation generally become manifest, depending on the nature of the product and especially on the protein and fat composition and content, at doses essentially above 2.5 kGy. These organoleptic side effects can be greatly limited when foods are irradiated in the chilled or, even better, frozen state, and particularly in the absence of oxygen.

In the past the approach has been taken that irradiation "adds" something to the treated food and that it should therefore be considered as a food additive and not as a process. The "food additive" approach to food irradiation meant that evaluation of the toxicological aspects of wholesomeness had to be based on the concepts of an acceptable daily intake and safety factors, as is the case with food additives or chemical residues in food. However, at present, it is recognized that the evaluation of the wholesomeness of irradiated foods more closely resembles a processing treatment and poses problems of a different kind from those encountered with food additives.

Another unproven suspicion was that irradiation might induce dangerous mutations of microbiological flora. However, this is unlikely unless an irradiated population of bacteria is repeatedly allowed to grow and is irradiated again to encourage such mutations. No known microbiological safety problems are produced when moist foods, such as fresh poultry, meat, and seafoods, are treated with low doses of ionizing energy, provided that the foods are properly refrigerated. This combined treatment not only extends the shelf-life but also reduces sensibly the hazard of disease-causing organisms, such as salmonellae, shigellae, coliforms, staphylococci, trichinae, *Yersinia enterocolitica*, *Campylobacyer jejuni*, and *Aeromonas hydrophila*. Refrigeration below 3 °C is needed to avoid the development of *Clostridium botulinum* type E if ionizing energy exceeding 1 kGy has been used to reduce the population of spoilage organisms.

The scientific aspects of the dispute on safety and wholesomeness of irradiated foods are detailed in the Federal Register (FDA 1986). The conclusions of these studies, that food irradiation is safe and has no serious harmful effects on the taste or wholesomeness of many foods, unless the levels of irradiation do not exceed prescribed limits, have provided guidelines for national authorities on clearing irradiated foods for human consumption. The official approval is either on an unconditional or provisional basis. An international logo to appear on irradiated foods consists of a solid

Fig. 7. The international logo for irradiated foods

circle, representing an energy source, above two petals, which represent the food. The five breaks in the outer circle depict rays from the energy source (Fig. 7).

Table 8 lists the countries that have cleared irradiated food for human consumption; this was compiled by the WHO in collaboration with the FAO of the United Nations.

1.6 Analytical Methods for Postirradiation Dosimetry of Foods

Legal acceptance and trade of irradiated foods require appropriate means of control, that is, a foolproof test to detect whether or not food has been irradiated, and eventually to quantify the amount of radiation. Such an assay is also relevant for avoiding repeated irradiation of the same food. Although the concentration of radiolytic products accumulated with repeated irradiation are generally so low that the potential toxicological hazard is minimal, the food is likely to be degraded in terms of organoleptic acceptability and, eventually, nutritional quality.

An ideal analytical method should be able to distinguish irradiated food in the absence of a control sample. The measured response after irradiation should be specific and proportional to the dose, and should be unaffected by various processing and storage conditions or length of time between irradiation treatment and analysis. In reality, since the irradiation process is remarkably gentle when compared with, for example, heat pasteurization, and the radiation-induced changes are generally very small and nonspecific, and because of natural variability of foods, a truly effective evaluation of radiation dose is exceedingly difficult. In foodstuffs, unique radiolytic products have not yet been isolated. Therefore, beyond the commercial and legislative interests, the development of methods that permit the identification of irradiated foods presents also a scientific challenge. The search for a suitable postirradiation detection method has been extensive (IAEA

Table 8. List of countries that have cleared irradiated food for human consumption (updated 22 March 1988)

Country/Product	Purpose of irradiation	Type of clearance	Dose permitted (kGy)	Date of approval
Argentina				
Strawberries	Shelf-life extension	Unconditional	2.5 max	30-4-1987
Potatoes	Sprout inhibition	Unconditional	0.03–0.15	30-4-1987
Onions	Sprout inhibition	Unconditional	0.02–0.15	30-4-1987
Garlic	Sprout inhibition	Unconditional	0.03–0.15	30-4-1987
Bangladesh				
Chicken	Shelf-life extension/ decontamination	Unconditional	Up to 8	28-12-1983
Papaya	Insect disinfestation/ control of ripening	Unconditional	Up to 1	28-12-1983
Potatoes	Sprout inhibition	Unconditional	Up to 0.15	28-12-1983
Wheat and ground wheat products	Insect disinfestation	Unconditional	Up to 1	28-12-1983
Fish	Shelf-life extension/ decontamination/ insect disinfestation	Unconditional	Up to 2.2	28-12-1983
Onions	Sprout inhibition	Unconditional	Up to 0.15	28-12-1983
Rice	Insect disinfestation	Unconditional	Up to 1	28-12-1983
Frog legs	Decontamination	Provisional		
Shrimp	Shelf-life extension/ decontamination	Provisional		
Mangos	Shelf-life extension/ insect disinfestation/ control of ripening	Unconditional	Up to 1	28-12-1983
Pulses	Insect disinfestation	Unconditional	Up to 1	28-12-1983
Spices	Decontamination/ insect disinfestation	Unconditional	Up to 10	28-12-1983

Table 8. *Continued*

42

Country/Product	Purpose of irradiation	Type of clearance	Dose permitted (kGy)	Date of approval
Belgium				
Potatoes	Sprout inhibition	Provisional	Up to 0.15	16-7-1980
Strawberries	Shelf-life extension	Provisional	Up to 3	16-7-1980
Onions	Sprout inhibition	Provisional	Up to 0.15	16-10-1980
Garlic	Sprout inhibition	Provisional	Up to 0.15	16-10-1980
Shallots	Sprout inhibition	Provisional	Up to 0.15	16-10-1980
Black, white pepper	Decontamination	Provisional	Up to 10	16-10-1980
Paprika powder	Decontamination	Provisional	Up to 10	16-10-1980
Arabic gum	Decontamination	Provisional	Up to 10	29-9-1983
Spices (178 different products)	Decontamination	Provisional	Up to 10	29-9-1983
Brazil				
Rice	Insect disinfestation	Unconditional	Up to 1	7-3-1985
Potatoes	Sprout inhibition	Unconditional	Up to 0.15	7-3-1985
Onions	Sprout inhibition	Unconditional	Up to 0.15	7-3-1985
Beans	Insect disinfestation	Unconditional	Up to 1	7-3-1985
Maize	Insect disinfestation	Unconditional	Up to 0.5	7-3-1985
Wheat	Insect disinfestation	Unconditional	Up to 1	7-3-1985
Wheat flour	Insect disinfestation	Unconditional	Up to 1	7-3-1985
Spices (13 different products)	Insect disinfestation/ decontamination	Unconditional	Up to 10	7-3-1985
Papaya	Insect disinfestation/ control of ripening	Unconditional	Up to 1	7-3-1985
Strawberries	Shelf-life extension	Unconditional	Up to 3	7-3-1985
Fish and fish products (fillets, salted, smoked dried, dehydrated)	Shelf-life extension/ decontamination/ insect disinfestation	Unconditional	Up to 2.2	8-3-1985
Poultry	Shelf-life extension/ decontamination	Unconditional	Up to 7	8-3-1985

Bulgaria

Potatoes	Sprout inhibition	Experimental batches	0.1	30-4-1972
Onions	Sprout inhibition	Experimental batches	0.1	30-4-1972
Garlic	Sprout inhibition	Experimental batches	0.1	30-4-1972
Grain	Insect disinfestation	Experimental batches	0.3	30-4-1972
Dry food concentrates	Insect disinfestation	Experimental batches	1	30-4-1972
Dried fruits	Insect disinfestation	Experimental batches	1	30-4-1972
Fresh fruits (tomatoes, peaches, apricots, cherries, raspberries, grapes)	Shelf-life extension	Experimental batches	2.5	30-4-1972

Canada

Potatoes	Sprout inhibition	Unconditional	Up to 0.1	9-11-1960 / 14-6-1963
Onions	Sprout inhibition	Unconditional	Up to 0.15	25-3-1965
Wheat, flour	Insect disinfestation	Unconditional	Up to 0.75	25-2-1969
Poultry	Decontamination	Test marketing	Up to 7	20-6-1973
Cod and haddock fillets	Shelf-life extension	Test marketing	Up to 1.5	2-10-1973
Spices and certain dried vegetables, seasonings	Decontamination	Unconditional	Up to 10	3-10-1984
Onion powder	Decontamination	Unconditional	Up to 10	12-12-1983

Chile

Potatoes	Sprout inhibition	Unconditional	Up to 0.15	29-12-1982
Papaya	Insect disinfestation	Unconditional	Up to 1	29-12-1982
Wheat and ground-wheat products	Insect disinfestation	Unconditional	Up to 1	29-12-1982
Strawberries	Shelf-life extension	Unconditional	Up to 3	29-12-1982
Chicken	Decontamination	Unconditional	Up to 7	29-12-1982
Onions	Sprout inhibition	Unconditional	Up to 0.15	29-12-1982
Rice	Insect disinfestation	Unconditional	Up to 1	29-12-1982
Teleost fish and fish products	Shelf-life extension/ decontamination/ insect disinfestation	Unconditional	Up to 2.2	29-12-1982

Table 8. *Continued*

Country/Product	Purpose of irradiation	Type of clearance	Dose permitted (kGy)	Date of approval
Cocoa beans	Decontamination/ insect disinfestation	Unconditional	Up to 5	29-12-1982
Dates	Insect disinfestation	Unconditional	Up to 1	29-12-1982
Mangos	Insect disinfestation/ shelf-life extension/ control of ripening	Unconditional	Up to 1	29-12-1982
pulses	Insect disinfestation	Unconditional	Up to 1	29-12-1982
Spices and condiments	Insect disinfestation/ Decontamination	Unconditional	Up to 10	29-12-1982
China				
Potatoes	Sprout inhibition	Unconditional	Up to 0.2	30-11-1984
Onions	Sprout inhibition	Unconditional	Up to 0.15	30-11-1984
Garlic	Sprout inhibition	Unconditional	Up to 0.1	30-11-1984
Peanuts	Insect disinfestation	Unconditional	Up to 0.4	30-11-1984
Grain	Insect disinfestation	Unconditional	Up to 0.45	30-11-1984
Mushrooms	Growth inhibition	Unconditional	Up to 1	30-11-1984
Sausage	Decontamination	Unconditional	Up to 8	30-11-1984
Czechoslovakia				
Potatoes	Sprout inhibition	Unconditional	Up to 0.1	26-11-1976
Onions	Sprout inhibition	Unconditional	Up to 0.08	26-11-1976
Mushrooms	Growth inhibition	Experimental	Up to 2	26-11-1976
Denmark				
Spices and herbs	Decontamination	Unconditional	Up to 15 max., Up to 10 average	23-12-1985

Finland				
Dry and dehydrated spices and herbs	Decontamination	Unconditional	Up to 10 average	13-11-1987
All foods for patients requiring a sterile diet	Sterilization	Unconditional	Unlimited	13-11-1987
France				
Potatoes	Sprout inhibition	Provisional	0.075–0.15	8-11-1972
Onions	Sprout inhibition	Provisional	0.075–0.15	9-8-1977
Garlic	Sprout inhibition	Provisional	0.075–0.15	9-8-1977
shallots	Sprout inhibition	Provisional	0.075–0.15	9-8-1977
Spices and aromatic substances (72 products)	Decontamination	Unconditional	Up to 11	10-2-1983
Gum arabic	Decontamination	Unconditional	Up to 9	16-6-1985
Muesli-like cereal	Decontamination	Unconditional	Up to 10	16-6-1985
Dehydrated vegetables	Decontamination	Unconditional	Up to 10	16-6-1985
Mechanically deboned poultry meat	Decontamination	Unconditional	Up to 5	16-2-1985
Dried fruits	Insect disinfestation	Unconditional	1 max.	6-1-1988
Dried vegetables	Insect disinfestation	Unconditional	1 max.	6-1-1988
Germany (GDR)				
Onions	Sprout inhibition	Unconditional	20	30-1-1984
Enzyme solutions	Decontamination	Unconditional	10	7-6-1983
Spices	Decontamination	Provisional	Up to 10	29-12-1982
Hungary				
Strawberries	Shelf-life extension	Test marketing		5-3-1973
Mixed dry ingredients for canned hashed meat	Decontamination	Experimental batches	5	20-11-1976
Onions (for dehydrated flakes' processing)	Sprout inhibition	Test marketing	0.05	18-11-1980

Table 8. *Continued*

46

Country/Product	Purpose of irradiation	Type of clearance	Dose permitted (kGy)	Date of approval
Mushrooms	Growth inhibition	Test marketing		15-4-1982
(*Agaricus*)			2.5	
(*Pleurotus*)			3	
Strawberries	Shelf-life extension	Test marketing	2.5	20-6-1981
Potatoes	Sprout inhibition	Test marketing	0.1	2-12-1981
Spices for sausage	Decontamination	Test marketing	5	28-6-1982
Grapes	Shelf-life extension	Test marketing	2.5	15-4-1982
Cherries	Shelf-life extension	Test marketing	2.5	15-4-1982
Sour cherries	Shelf-life extension	Test marketing	2.5	15-4-1982
Red currants	Shelf-life extension	Test marketing	2.5	15-4-1982
Onions	Sprout inhibition	Unconditional	0.05–0.02	23-6-1982
Pears	Shelf-life extension	Test marketing	2.5	7-12-1982
Pears	Shelf-life extension	Test marketing	1.0 + CaCl$_2$ treatment	24-1-1983
Potatoes (for processing into flakes)	Sprout inhibition	Test marketing	0.1	28-1-1983
Frozen chicken	Decontamination	Test marketing	4	3-10-1983
Sour cherries (canned)		Conditional	0.2 average	23-4-1985
Black pepper	Decontamination	Conditional	6 minimum	5-1985
Spcies	Decontamination	Unconditional	8, 6 average	19-8-1986
India				
Potatoes	Sprout inhibition	Unconditional	Codex standard	1-1986
Onions	Sprout inhibition	Unconditional	Codex standard	1-1986
Spices	Disinfection	For export only	Codex standard	1-1986
Frozen shrimps and frog legs	Disinfection	For export only	Codex standard	1-1986
Indonesia				
Tuber and root crops (potatoes, shallots, garlic and rhizomes)	Sprout inhibition	Unconditional	0.15 max.	29-12-1987

Analytical Methods for Postirradiation Dosimetry of Foods

Cereals	Disinfestation	Unconditional	1 max.	29-12-1987
Dried Spices	Decontamination	Unconditional	10 max.	29-12-1987
Israel				
Potatoes	Sprout inhibition	Unconditional	0.15 max.	5-7-1967
Onions	Sprout inhibition	Unconditional	0.10 max.	25-7-1968
Onions	Sprout inhibition	Unconditional	0.15	6-3-1985
Garlic	Sprout inhibition	Unconditional	0.15	6-3-1985
Shallots	Sprout inhibition	Unconditional	0.15	6-3-1985
Spices (36 different products)	Decontamination	Unconditional	10	6-3-1985
Fresh fruits and vegetables grains, cereals, pulses, cacao and coffee beans, nuts, edible seeds	Disinfestation	Unconditional	1 average	1-1987
Mushrooms, strawberries	Shelf-life extension	Unconditional	3 average	1-1987
Poultry and poultry sections	Decontamination	Unconditional	7 average	1-1987
Spices and condiments, dehydrated and dried vegetables, edible herbs	Decontamination	Unconditional	10 average	1-1987
Poultry feeds	Decontamination	Unconditional	15 average	1-1987
Italy				
Potatoes	Sprout inhibition	Unconditional	0.075–0.15	30-8-1973
Onions	Sprout inhibition	Unconditional	0.075–0.15	30-8-1973
Garlic	Sprout inhibition	Unconditional	0.075–0.15	30-8-1973
Japan				
Potatoes	Sprout inhibition	Unconditional	0.15 max.	30-8-1972
Korea				
Potatoes	Sprout inhibition	Unconditional	0.15 max.	28-9-1987
Onions	Sprout inhibition	Unconditional	0.15 max.	28-9-1987
Garlic	Sprout inhibition	Unconditional	0.15 max.	28-9-1987

Table 8. *Continued*

48

Country/Product	Purpose of irradiation	Type of clearance	Dose permitted (kGy)	Date of approval
Chestnuts	Sprout inhibition	Unconditional	0.25 max.	28-9-1987
Fresh and dry mushrooms	Growth inhibition/ insect disinfestation	Unconditional	1 max.	28-9-1987
Deep frozen meals	Sterilization	Hospital patients	25 min.	27-11-1969
Potatoes	Sprout inhibition	Unconditional	0.15 max.	23-3-1970
Shrimps	Shelf-life extension	Experimental batches	0.5–1	13-11-1970
Onions	Sprout inhibition	Experimental batches	0.15	5-2-1971
Spices and condiments	Decontamination	Experimental batches	8–10	13-9-1971
Poultry, eviscerated (in plastic bags)	Shelf-life extension	Experimental batches	3 max.	31-12-1971
Chicken	Shelf-life extension/ decontamination	Unconditional	3 max.	10-5-1976
Fresh, tinned and liquid foodstuffs	Sterilization	Hospital patients	25 min.	8-3-1972
Powdered batter mix	Decontamination	Test marketing	1.5	4-10-1974
Vegetable filling	Decontamination	Test marketing	0.75	4-10-1974
Endive (prepared, cut)	Shelf-life extension	Test marketing	1	14-1-1975
Onions	Sprout inhibition	Unconditional	0.05 max.	9-6-1975
Spices	Decontamination	Provisional	10	26-6-1975
Peeled potatoes	Shelf-life extension	Test marketing	0.5	12-5-1976
Chicken	Shelf-life extension/ decontamination	Unconditional	3 max.	10-5-1976
Shrimps	Shelf-life extension	Test marketing	1	15-6-1976
Fillets of haddock, coal-fish, whiting	Shelf-life extension	Test marketing	1	6-9-1976
Fillets of cod and plaice	Shelf-life extension	Test marketing	1	7-9-1976
Fresh vegetables (prepared, cut, soup greens)	Shelf-life extension	Test marketing	1	6-9-1977
Frozen frog legs	Decontamination	Provisional	5	25-9-1978

Rice and ground rice products	Insect disinfestation	Provisional	1	15-3-1979
Rye bread	Shelf-life extension	Provisional	5 max.	12-2-1980
Spices	Decontamination	Provisional	7 max.	15-4-1980
Malt	Decontamination	Provisional	10 max.	8-2-1983
Boiled and cooled shrimp	Shelf-life extension	Provisional	1 max.	8-2-1983
Frozen shrimp	Decontamination	Provisional	7 max.	8-2-1983
Frozen fish	Decontamination	Provisional	6 max.	24-8-1983
Egg powder	Decontamination	Provisional	6 max.	25-8-1983
Dry blood protein	Decontamination	Provisional	7 max.	25-8-1983
Dehydrated vegetables	Decontamination	Provisional	10 max.	27-10-1983
Refrigerated snacks of minced meat	Shelf-life extension	Test marketing	2	12-7-1984
New Zealand				
Herbs and spices (one batch)	Decontamination	Provisional	8	3-1985
Norway				
Spices	Decontamination	Unconditional	Up to 10	
Philippines				
Potatoes	Sprout inhibition	Provisional	0.15 max.	13-9-1972
Onions	Sprout inhibition	Provisional	0.07	1981
Garlic	Sprout inhibition	Provisional	0.07	1981
Onions and garlic	Sprout inhibition	Test marketing		29-9-1986
Poland				
Potatoes	Sprout inhibition	Provisional	Up to 0.15	1982
Onions	Sprout inhibition	Provisional		3-1983
South Africa				
Potatoes	Sprout inhibition	Unconditional	0.12–0.24	19-1-1977
Dried bananas	Insect disinfestation	Provisional	0.5 max.	28-7-1977

Table 8. *Continued*

Country/Product	Purpose of irradiation	Type of clearance	Dose permitted (kGy)	Date of approval
Avocados	Insect disinfestation	Provisional	0.1 max.	28-7-1977
Onions	Sprout inhibition	Unconditional	0.05–0.15	25-8-1978
Garlic	Sprout inhibition	Unconditional	0.1–0.20	25-8-1978
Chicken	Shelf-life extension/ decontamination	Unconditional	2–7	25-8-1978
Papaya	Shelf-life extension	Unconditional	0.5–1.5	25-8-1978
Mango	Shelf-life extension	Unconditional	0.5–1.5	25-8-1978
Strawberries	Shelf-life extension	Unconditional	1–4	25-8-1978
Bananas	Shelf-life extension	Unconditional		1982
Litchis	Shelf-life extension	Unconditional		1982
Pickled mango (achar)	Shelf-life extension	Unconditional		1982
Avocados	Shelf-life extension	Unconditional		1982
Frozen fruit juices	Shelf-life extension	Unconditional		
Green beans		Unconditional		
Tomatoes	Control of ripening	Unconditional		
Brinjals		Unconditional		
Soya pickle products		Unconditional		
Ginger		Unconditional		
Vegetable paste		Unconditional		
Bananas (dried)	Insect disinfestation	Unconditional		
Almonds	Insect disinfestation	Unconditional		
Cheese powder	Insect disinfestation	Unconditional		
Yeast powder		Unconditional		
Herbal tea		Unconditional		
Various spices		Unconditional		
Various dehydrated vegetables		Unconditional		

Spain

Potatoes	Sprout inhibition	Unconditional	0.05–0.15	4-11-1969
Onions	Sprout inhibition	Unconditional	0.08 max.	1971

Thailand

Onions	Sprout inhibition	Unconditional	0.1 max.	20-3-1973
Potatoes, onions, garlic	Sprout inhibition	Unconditional	0.15	4-12-1986
Dates	Disinfestation/Delay of ripening	Unconditional	1	4-12-1986
Mangos, papaya		Unconditional	1	4-12-1986
Wheat, rice, pulses	Disinfestation	Unconditional	1	4-12-1986
Cacao beans	Disinfestation	Unconditional	1	4-12-1986
Fish and fishery products	Disinfestation	Unconditional	1	4-12-1986
Fish and fishery products	Reduce microbial load	Unconditional	2.2	4-12-1986
Strawberries	Shelf-life extension	Unconditional	3	4-12-1986
Nam	Decontamination	Unconditional	4	4-12-1986
Moo yor	Decontamination	Unconditional	5	4-12-1986
Sausage	Decontamination	Unconditional	5	4-12-1986
Frozen shrimps	Decontamination	Unconditional	5	4-12-1986
Cacao beans	Reduce microbial load	Unconditional	5	4-12-1986
Chicken	Disinfestation/ shelf-life extension	Unconditional	7	4-12-1986
Spices and condiments	Insect disinfestation	Unconditional	1	4-12-1986
Dehydrated onions and onions powder	Decontamination	Unconditional	10	4-12-1986

USSR

Potatoes	Sprout inhibition	Unconditional	0.3	17-7-1973
Grain	Insect disinfestation	Unconditional	0.3	1959
Fresh fruits and vegetables	Shelf-life extension	Experimental batches	2–4	11-7-1964

Table 8. *Continued*

Country/Product	Purpose of irradiation	Type of clearance	Dose permitted (kGy)	Date of approval
Semi prepared raw beef, pork and rabbit products (in plastic bags)	Shelf-life extension	Experimental batches	6–8	11-7-1964
Dry food concentrates (buckwheat mush, gruel, rice, puddings)	Insect disinfestation	Unconditional	0.7	6-6-1966
Poultry, eviscerated (in plastic bags)	Shelf-life extension	Experimental batches	6	4-7-1966
Culinary prepared meat products (fried meat, entrecote) (in plastic bags)	Shelf-life extension	Test marketing	8	1-2-1967
Onions	Sprout inhibition	Unconditional	0.06	17-7-1973
United Kingdom				
Any food for consumption by patients who require a sterile diet as an essential factor in their treatment	Sterilization		Hospital patients	1-12-1969
USA				
Wheat and wheat flour	Insect disinfestation	Unconditional	0.2–0.5	21-8-1963
White potatoes	Shelf-life extension	Unconditional	0.05–0.15	1-11-1965
Spices and dry vegetable seasonings (38 commodities)	Decontamination/ Insect disinfestation		30 max.	5-7-1983
Dry or dehydrated enzymes preparations (including immobilized enzyme preparation)	Control of insects and/or microorganisms	Unconditional	10 max.	10-6-1985

Analytical Methods for Postirradiation Dosimetry of Foods

Food	Purpose of process	Condition	Dose (kGy)	Date
Pork carcasses or fresh, non-heat processed cuts of pork carcasses	Control of *Trichinella spiralis*	Unconditional	0.3 min.–1.0 max.	22-7-1985
Fresh foods	Delay or maturation	Unconditional	1	18-4-1986
Food	Disinfestation	Unconditional	1	18-4-1986
Dry or dehydrated enzyme preparations	Decontamination	Unconditional	10	18-4-1986
Dry or dehydrated aromatic vegetable substances	Decontamination	Unconditional	30	18-4-1986
Uruguay				
Potatoes	Sprout inhibition	Unconditional		23-6-1970
Yugoslavia				
Cereals	Insect disinfestation	Unconditional	Up to 10	17-12-1984
Legumes	Insect disinfestation	Unconditional	Up to 10	17-12-1984
Onions	Sprout inhibition	Unconditional	Up to 10	17-12-1984
Garlic	Sprout inhibition	Unconditional	Up to 10	17-12-1984
Potatoes	Sprout inhibition	Unconditional	Up to 10	17-12-1984
Dehydrated fruits and vegetables	Sprout inhibition	Unconditional	Up to 10	17-12-1984
Dried mushrooms	Decontamination	Unconditional	Up to 10	17-12-1984
Egg powder	Decontamination	Unconditional	Up to 10	17-12-1984
Herbal teas, tea extracts	Decontamination	Unconditional	Up to 10	17-12-1984
Fresh poultry	Decontamination/ Shelf-life extension	Unconditional	Up to 10	17-12-1984

Recommendations published by international organizations

FAO/IAEA/WHO Expert Committee 1976

Food	Purpose of process	Condition	Dose (kGy)	Date
Potatoes	Sprout inhibition	Unconditional	0.03–0.15	7-9-1976
Onions	Sprout inhibition	Provisional	0.02–0.15	7-9-1976
Papaya	Insect disinfestation	Unconditional	0.5–1	7-9-1976

Table 8. *Continued*

Country/Product	Purpose of irradiation	Type of clearance	Dose permitted (kGy)	Date of approval
Strawberries	Shelf-life extension	Unconditional	1–3	7-9-1976
Wheat and ground-wheat products	Insect disinfestation	Unconditional	0.15–1	7-9-1976
Rice	Insect disinfestation	Provisional	0.1–1	7-9-1976
Chicken	Shelf-life extension/ decontamination	Unconditional	2–7	7-9-1976
Cod and redfish	Shelf-life extension/ decontamination	Provisional	2–2.2	7-9-1976
FAO/IAEA/WHO Expert Committee 1980				
Any food product	Sprout inhibition/ shelf-life extension/ decontamination/ insect disinfestation/ control of ripening/growth inhibition	Unconditional	Up to 10	3-11-1980

Measurements of Physical Effects 55

1991, Raffi and Belliardo 1991). The evaluated methods can be classified in three groups: (1) Measurements of physical effects; (2) measurements of chemical effects; (3) microbiological and biological methods.

1.6.1 Measurements of Physical Effects

Direct Detection of Free Radicals. The absorption of ionizing radiation by food molecules leads to formation of free radicals. Radicals are generally formed as a direct result of radiolysis, dissociation of radical cations, reactions between electrons and molecules, and also as a consequence of ion–molecule reactions:

$$RH \rightarrow R\cdot + H\cdot \tag{1}$$

$$RH^+ + RH \rightarrow R\cdot + RH_2^+ \tag{2}$$

$$A + e^- \rightarrow A^- \tag{3}$$

$$RX + e^- \rightarrow R\cdot + X^-. \tag{4}$$

The momentary concentration of free radicals depends on the nature of the material irradiated, the radiation dose, and the time interval between irradiation treatment and radical measurement.

If the process of radical formation occurs in a constrained matrix, such as polycrystalline or amorphous material, the original free radicals may persist, may reform the original bond, or may react to form secondary long-lived radicals. Since the ultimate fate of the radicals depends on their ability to move into a position for bimolecular reaction, factors that alter their environment, such as water activity, predetermine the pathways for decay and hence, the lifetime. Indeed, the half-life of free radicals produced in irradiated solid foods decreased with an increase in the humidity of the system (O'Meara and Shaw 1957; Fritsch and Reymond 1970; Diehl 1972; Uchiyama and Uchiyama 1979; Raffi and Agnel 1983).

The broad electron paramagnetic resonance (EPR) signals of free radicals trapped in a solid matrix are poorly resolved and do not allow chemical assignments, but the EPR determination of total free radical content in dry materials, like spices, could serve as an indicative analytical tool to determine whether or not the food has been irradiated. The intensity of the EPR signal of irradiated spices is approximately proportional to the total irradiation dose, between the background signal and the saturation dose. This method is limited by the background variation in the radical content, which requires the analysis of a control nonirradiated sample, and the gradual postirradiation loss of signal with storage time (Yang et al. 1987; Davidson and Forrester 1988; Troup et al. 1989). The free radicals in dry foods cannot be defined as "unique radiolytic products".

Nonirradiated spices, vanilla beans, and heated cereal grains contain stable paramagnetic species which apparently originate in the phenolic constituents of the vegetable material. The preirradiation free radical content of vegetable foods varies substantially, due, presumably, to the different processing conditions: grinding, exposure to sunlight, or heat drying.

Raffi et al. (1988) identified a signal in the EPR spectrum which was only present in the achenes of irradiated strawberries. Within the limits of commercial handling of strawberries, e.g., less than 20 days of storage at 5 °C, the EPR test was able to identify all samples irradiated by a dose of at least 1 kGy. Desrosiers and McLauglin (1989) observed an EPR signal in mango seeds which decayed after a few days.

Some hard, calcified tissues that occur in foods like pork, chicken and cod bones or prawn cuticles also stabilize free radicals (Desrosiers and Simic 1988; Desrosiers 1989; Raffi et al. 1989). The signal, which is very long lived and even survives cooking, probably originates in the hydroxyapatite content of the bone (Dodd et al. 1988, 1989; Lea et al. 1988; Gray and Stevenson 1989a,b; Stevenson and Gray 1989a,b; Gray et al. 1990). The presence of such a signal can provide evidence of irradiation with doses as low as 200 Gy. The radiation-absorbed dose given to such foods could be estimated by the additive reirradiation of bone samples, followed by the backward extrapolation of the EPR signal against dose fraction. This approach potentially eliminates the need for examination of nonirradiated samples (Desrosiers 1990).

Lyoluminescence (Chemiluminescence). The term "lyoluminescence" applies to emission of light on dissolution of a solid in a liquid. One of the most common lyoluminescence effects is when solid samples of various substances irradiated with ionizing radiation are dissolved or brought into contact with a solvent, generally water. The integral light yield increases with the administered radiation dose up to a saturation value. The ubiquity of lyoluminescence of irradiated solids is well attested. Spices, dried foods like powdered milk and soups, cotton, cloth, and paper generate light when placed in contact with a solvent (Ettinger and Puite 1982). Although one common mechanism is insufficient to explain this effect in such a broad variety of materials, there is a common denominator for the chemical basis of lyoluminescence – namely the production and subsequent reactions of free radicals. Radicals formed are trapped in the solid for relatively long times, particularly at lower temperatures. In a liquid medium, free radical reactions can take place and liberate enough energh energy for light emission in the short wavelength part of the visible spectrum. The lyoluminescence yield can be increased by adding chemiluminescent sensitizers, such as the combination of luminol and hemin, adjusted to pH 10–11 (λ_{max} approx. 424 nm). Although the exact mechanism of activation of chemiluminescence of irradiated solids by luminol has not been thoroughly explained, the free radicals are the initiators of the sequence of processes leading to intensely luminescent reactions.

A variety of spices, milk powder, whole onions, and frozen chicken (particularly cartilage tissue rather than bone or meat) irradiated with doses up to 10 kGy showed lyoluminescence in a luminol solution (Bogl and Heide 1985). Likewise, gamma-irradiated pepper could be detected (Sattar et al. 1987). In some foods it was possible to identify the radiation treatment as late as several months after irradiation. The method is, however, plagued by problems including fading of the chemiluminescent response with storage time and poor reproducibility.

Thermoluminescence. The heat-stimulated emission of light (thermoluminescence) from irradiated solids has also been evaluated as a method for identifying irradiated foods. Irradiated herbs, spices, and seasonings exposed to gamma-ray doses from 1 to 20 kGy emit light when heated from room temperature to 400–500 °C with a linear heating rate of 6–8 °C/s and in a dry nitrogen atmosphere. Today, it is generally recognized that the thermoluminescence signals usually originate from the minute amount of mineral dust adhering to the sample surface (Goksu-Ogelman and Regulla 1989; Sanderson et al. 1989a,b). Dating studies have shown that these minerals have thermoluminescence abilities ranging from 10^3 to 10^6 photons $s^{-1} mg^{-1} Gy^{-1}$ and, therefore, microgram quantities can easily yield observable signals. Sensitivity enhancements of 10^3 or more, and improved signal to blank ratios, were recorded for measurements of the inorganic dust separated from the food. On the basis of isothermal decay experiments, the half-life of the signals was estimated to be about 5 years at room temperature. The measurements on separated mineral grains seem to allow reliable identification of irradiated spices with 90% confidence.

The thermoluminescence of irradiated minerals has been known of for a long time. Robert Boyle observed this property in crystals in 1634. Boyle described experiments he carried out on a diamond which belonged to a Mr. Clayton. Boyle wrote ". . . I also brought it to some kind of Glimmering Light, by taking it into bed with me, and holding it a good while upon a warm part of my naked body." Although Boyle found that he could not elicit light from other precious stones by taking them to bed with him, modern sensitive detection systems can measure thermoluminescence from most natural inorganic materials. The phenomenon is due to irradiation from radioactive elements such as thorium, uranium, and potassium-40, which are present as trace elements in most rocks and soils (Bishop 1988).

It can be expected that the use of more refined thermoluminescence evaluation methods, resolving the different thermoluminescent components and discriminating those without adequate dosimetric properties, will result in a more reliable tool of thermoluminescence for detection and evaluation of irradiation doses in foods.

58   Ionizing Radiation

Other Physical Measurements. Electrical conductivity measurements seem to be a rapid method for identification of irradiated potatoes. Using this method nonirradiated and irradiated potatoes could not only be differentiated but also the applied dose could be estimated after storage of the potatoes for up to 6 months. For different potato varieties a probability of at least 75% was achieved in detecting radiation processing. The biological-physicochemical effect underlying the method has not been explained. Electrical conductivity measurements of other vegetables showed no consistent alteration by irradiation (Hayashi and Kawashima 1983). Electrical resistance measurements were suggested for identification of irradiated fish (Ehlermann 1972).

Viscosity measurements have shown some promise (Farkas et al. 1990), though the effects observed are not understood and nor were they consistent from one product to another, i.e., in some materials the viscosity decreased after irradiation but in other materials it increased. There is no correlation between the observed effects and the starch content of the food.

Differential scanning calorimetry revealed changes in the initial freezing point or in the heterogeneous nucleation temperature of food products submitted to irradiation.

1.6.2 Measurements of Chemical Effects

Detection of Products Derived from Radiolysis of Lipids. The major radiolytic hydrocarbons resulting from specific cleavage of fatty acid residues in triglycerides can be used as indicators of irradiation in fatty foods. The radiolytic splitting of fat does not occur randomly; two hydrocarbons are produced preferentially from each fatty acid. One, resulting from preferential cleavage at the carbon–carbon bond alpha to the carbonyl group, has a carbon atom less than the parent fatty acid. The other has two carbon atoms less than the parent fatty acid, has an extra double bond, and results from cleavage at the carbon–carbon bond beta to the carbonyl group. These major radiolytic hydrocarbons reflect the severity of the irradiation treatment, since their production increases linearly with dose and temperature of irradiation. The presence of moisture or air during irradiation does not significantly alter the radiolytic pattern. Experiments with ground pork indicate that these compounds are present in samples irradiated at doses as low as 1 kGy but are absent in nonirradiated or heated samples (Nawar and Balboni 1970; Vajdi and Nawar 1979; Vajdi and Merritt 1985; Meier and Biedermann 1990). Volatile hydrocarbons and aldehydes were detected in irradiated chicken meat using on-line coupled liquid chromatography – gas chromatography (Meier et al. 1990) and in irradiated frog legs by a gas chromatography method (Morehouse and Ku

Measurements of Chemical Effects 59

1990). The estimations of the applied dose for frog legs, of unknown origin, by gas chromatography data on the hydrocarbons formed during radiolysis of lipids and by ESR measurements of free radicals trapped in the bone were in good agreement (Morehouse et al. 1991).

Irradiation of simple triglycerides yields also substituted cyclobutanones with an alkyl group at ring position 2 and with as many carbon atoms as the parent fatty acid. The substance 2-dodecyl cyclobutanone, formed from palmitic acid, was suggested as a potential postirradiation marker for minced chicken meat. The concentration of 2-dodecyl cyclobutanone produced by 5 kGy irradiation is approximately $0.2 \mu g/g$ fresh meat weight, five times the estimated limit of detection (Stevenson et al. 1990). The compound was not detected in either raw or cooked nonirradiated minced chicken meat and it was detectable for 20 days postirradiation (Boyd et al. 1991).

Radiation-induced oxidation of lipids can also, a priori, be a suitable reaction for detection of irradiated food because of the amplifying effect achieved by the chain character of this reaction. Indeed, the lipid hydroperoxide content has been proposed as an indicator of the irradiation of egg and milk powders and soya flour. The formation of hydroperoxides appears to depend only on the radiation dose and even after 6 months the level was still higher than the background value of nonirradiated samples (Katusin-Razem et al. 1990). However, a few parameters unrelated to radiation, like temperature, exposure to light, the availability of oxygen, and traces of catalytic metal ions, may affect the extent of lipid oxidation. A recent study on cholesterol oxidation indicated that the change in the ratio of 7-ketocholesterol to the sum of cholesterol 5α-epoxide and $5\varphi, 6\varphi$-epoxide may be a means of determining whether or not meat or other foods containing cholesterol have been subjected to ionizing radiation (Lakritz and Maerker 1989). However, results with beef indicated that the content of cholesterol oxidation products was more affected by the origin of the sample than by irradiation (Zabielski 1989).

Radiolytic Hydroxylation of Proteins. This analytical approach is primarily aimed at foods like raw meat which are rich in protein and contain more than 50% water. When exposed to ionizing radiation, the hydroxyl radical generated radiolytically from water interacts with amino acids found in the proteins. Among the resulting hydroxylated products, 2-hydroxyphenylalanine (*o*-tyrosine) is not naturally incorporated in proteins and can serve as an internal dosimeter. The formation of this compound can be monitored, after drying and hydrolysis of the meat sample, by gas chromatography – mass spectrometry (Karam and Simic 1988) or by high performance liquid chromatography with fluorescence detection (Meier et al. 1989). Because the formation of *o*-tyrosine depends not only on the dose but also on the

60 Ionizing Radiation

dose rate and the temperature during irradiation, it is not possible to determine the irradiation dose. It seems, however, that the major problem with *o*-tyrosine has been the variable levels of *o*-tyrosine found in non-irradiated foods (Hart et al. 1988).

"Unnatural" hydroxylated aromatic compounds generated after irradiation can also be readily separated from other phenolic constituents present in extracts of vegetable foodstuffs. Reverse-phase HPLC and an electrochemical detector equipped with a glassy carbon working electrode operating in oxidation mode were employed for this separation (Grootveld and Jain 1989b; Grootveld et al. 1990).

Release of Hydrogen Gas (Dohmaru et al. 1989). The method is based on the fact that hydrogen gas is formed in organic substances irradiated with ionizing radiation. Following gamma irradiation, black and white peppers were ground to a powder in a ceramic gastight mill and the released gas was analyzed for H_2 by gas chromatography. Identification of irradiated pepper was possible up to 4 months after 10 kGy irradiation.

Changes in DNA. The effects of radiation on genetic material, and DNA in particular, have been thoroughly investigated primarily because of their importance in radiation biology (von Sonntag 1987). DNA molecules are very sensitive to gamma radiation even at very low doses because of damage due to the formation of radiolytic products of bases or sugars, as well as structural changes due to strand breaks or crosslinking between bases and protein. Since very sensitive techniques for detection of radiation-induced lesions in nucleic acids are already available, and most foods derived from living organisms, meat, fish, and vegetables, always have DNA as a trace constituent, the use of DNA modifications for detection is most promising. Thus, the detection of the abnormal bases created in DNA by hydroxyl radicals such 8-hydroxyguanine and 5-hydroxycytosine have been suggested as a diagnostic test (Grootveld et al. 1990). The base products, however, are not strictly a result of irradiation. Hydroxyl radicals arise also in biochemical oxidation reactions and may cause similar damage. Unfortunately, neither are conformational changes, like strand breaks, specific to radiation; they are greatly influenced by storage and processing conditions (freezing, thawing, etc.). Background levels of nonirradiated food items, therefore, have to be checked carefully. There is a possibility that mitochondrial DNA is better protected from enzymatic reactions than cellular DNA, so its alterations might be more specific to irradiation.

Due to the rapid progress in molecular biology and gene engineering, methods of analyzing nucleic acids are gaining in sensitivity. It is therefore conceivable that if radiation-specific changes in DNA occur, it will be possible to develop an analytical detection procedure based on these changes.

Microbiological and Biological Methods 61

Detection of Radiolytic Products of Carbohydrates. Monosaccharides and polysaccharides are common components of various foodstuffs and are modified in specific patterns by radiolytically generated transients like hydroxyl radicals (den Drijver et al. 1986). Detection of oligosaccharide fragments derived from degradation of polysaccharides may have an application as an analytical test for certain types of irradiated foods. For example, the detection of N-acetylglucosamine oligosaccharides derived from the fragmentation of chitin, if their formation is radiolytically specific, may have potential in determining whether prawns or shrimps have been irradiated. The detection can be technically achieved by high-field proton Hahn spin-echo NMR spectroscopy or alternative techniques (Grootveld and Jain 1989a,b; Grootveld et al. 1990).

1.6.3 Microbiological and Biological Methods

Microbiological Methods. Microorganisms differ greatly, with species and even with strain, in their sensitivity to radiation. For example, Gram-negative bacteria are generally more sensitive than Gram-positive. Selective destruction of microorganisms, due to the different radioresistances, is expected in irradiated food. For example, the microflora on raw poultry meat shows a characteristic microbiological profile with significant numbers of Gram-negative bacteria, predominantly of the genus *Pseudomonas*. By contrast, the flora which develop on a raw chicken after irradiation at a dose of 2.5 kGy are mostly Gram-positive bacteria and yeasts. The presence of Gram-negative bacteria in both the live and dead states can be measured using the *Limulus* amoebocyte lysate (LAL) test, which is specific to these organisms, and the enumeration of the live cells can be achieved by counting colonies which grow on a selective agar. Taken together, a high LAL titre and a low Gram-negative bacterial count indicate irradiation treatment (McWeeny et al. 1991). A similar approach utilizes a comparison of an aerobic plate count with the direct epifluorescent filter technique. The latter technique counts organisms differentially stained by acridine orange irrespective of viability, and the difference to an aerobic plate count gives the number of organisms rendered nonviable by irradiation (Betts et al. 1988). The microbial characterization is not, however, confirmatory for irradiation since the shift in the microbial composition could be influenced by other factors, like the use of some preservatives, heat treatment, storage conditions, and environmental and growing factors for fruits. On the other hand, an advantage of the microbial method is that it provides additional information about the hygiene quality of the food.

The study of the microbial profile and the enzyme activity pattern of fresh food (uncooked) has been proposed in the past as a possible method to demonstrate undeclared application of radiation. The conceptual basis of this approach is that enzymes are usually not inactivated by usual doses of radiation and remain active during storage. The enzyme activity pattern of the treated food will be therefore be very similar to that of the raw food, while the microbial profile with probably be different from the one expected in a raw food.

Other Biological Methods Specific for Identification of Foods of Plant Origin. Several biological and histochemical measurements have been suggested for specific identification of irradiated potatoes (Thomas 1983) and onions (Thomas 1984). Some biological methods make use of the effect for which irradiation is used. Sprout inhibition of potatoes by irradiation is irreversible and may serve as a proof of irradiation, but the method is too slow for routine analysis (even if growth hormones are used to accelerate sprouting). The germination of half-embryo in grapefruits irradiated with doses above 0.15 kGy showed markedly reduced root growth and shoot elongation (Kawamura et al. 1989a,b). Enzymatic changes in irradiated potatoes could be histochemically visualized for a few weeks with a tetrazolium stain (Jona and Fonda 1990).

1.6.4 Conclusions

The lack of an absolute irradiation test is a contradictory reality. On the one hand, the inability to detect irradiated food is the strongest proof that the food is safe, and on the other hand, the absence of a surveillance test impedes acceptance of irradiation treatment.

Foodstuffs have different chemical compositions and physical states; the irradiation doses which can be applied, alone or combined with thermal treatments, vary in intensity according to the desired effect. Even within the same kind of food there are differences due to biological variability, growth conditions, and state of maturity, or differences caused by processing methods and storage. Therefore, it is unlikely that one method can be universally applicable for all irradiated foods. It becomes more and more clear that only a combination of analytical methods can solve the problem of detection, both from scientific and practical points of view. A rapid screening method, which is of low cost and relatively undemanding in skills and facilities, should be followed by a more refined, reliable confirmatory test, even if it is more time consuming and demanding of specialized skills and facilities. Modern methods of multicomponent analysis combined with multivariate statistical evaluation might be the solution to this complex problem.

1.7 Consumer Acceptance of Food Irradiation

The thorough studies of effects of ionizing radiation on foods have clarified virtually every technical aspect, leaving unanswered only the question of whether the public is willing to buy irradiated food and pay the extra cost for this treatment.

In spite of the great deal of accumulated information, the topic of food irradiation is still treated emotively. The opposition to food irradiation is caused by consumer sensitivity to new and unknown risks and by a natural desire for absolute certainty. In the course of time there have been many ill-founded claims and misunderstandings about the use of ionizing radiation and a lot of controversial information has been published, but the professional view is that the benefits far outweigh the risks. The commonly employed arguments on either side of the fence can be summarized as follows:

Pro

1. No competitive alternative for cleaning produce such as spices and tropical fruits.
2. The agents which cause spoilage (bacteria, insects) are eliminated. Irradiation of food has a particular bearing when hygiene is difficult to maintain in areas where food is being processed or when a perishable food has to be transported over long distances. Irradiation prolongs the shelf-life of a wide variety of foods. It can replace or reduce the use of chemical preservatives and protectors for food, so that there are no chemical residues in food.
3. Because the radiation does not heat the treated material, the food (fish, fruits, and vegetables) keeps its freshness and retains its physical state (frozen or dried commodities).
4. Food waste is reduced and this is particularly important to meet demands in less developed parts of the world.
5. An educated public will enjoy better and safer foods.

Against

1. Irradiation enables lax food hygiene by make unwholesome food appear fit for use.
2. While killing bacteria such as *Salmonella*, *Campylobacter*, and *Pseudomonas* the radiation will not remove toxins that have been created by bacteria in the earliest stages of contamination. *Clostridium* spores are not killed by irradiation and the organism could flourish without the warning odor of yeast and molds.
3. Nutrients, like vitamins, are lost. Although these losses are comparable to those occurring in cooking, this is not the case for foods that are eaten

fresh and often raw, such as fruits and vegetables. Food wholesomeness (flavor, odor) may be affected.
4. Building irradiation plants, roads, etc. is uneconomical in the developing world, where the problems of food preservation are most acute.
5. The inability to detect unique radiolytic products is a matter of limitation of the present analytical techniques. The degree of concern should not be dictated by a temporary state-of-the-art.
6. Psychological hostility to radiation exists.

At this stage of the book, the educated reader should be equipped to reach a reasonable position in this dispute.

Chapter 2   Ultraviolet-Visible Radiation

2.1 The Definition of Ultraviolet-Visible Radiation

The existence of ultraviolet irradiation in the solar spectrum was revealed in 1801, when a Jena physician, Johann Ritter (Ritter 1801), investigated the effectiveness of different colors of light in blackening crystals of silver chloride; he showed that radiation beyond the violet was most effective. The ultraviolet portion of the electromagnetic spectrum is divided between the radiation absorbed by the air, and therefore named "vacuum" (or "far") ultraviolet (10–200 nm), and the air-penetrating radiation of wavelengths of 200–380 nm. The latter range is subsegmented below and above *ca*. 300 nm. This value demarcates the upper limit for the solar wavelengths absorbed and filtered by the stratospheric ozone layer, meaning that only longer wavelengths reach the earth. It also differentiates between ultraviolet radiation of wavelengths shorter than 300 nm, which is absorbed by the fundamental components of the living cell, such as proteins and nucleic acids, and the radiation above this value which is not and is therefore less noxious.

The radiation of wavelengths between 380 and 780 nm is perceptible by the human eye and is, not surprisingly, defined as the visible portion of the spectrum. There is no fundamental physical difference between ultraviolet and visible radiations and the difference in nomenclature is justified only by the physiological limitation of the human eye. Seemingly, this is the reason that the term "light", although inaccurate, is liberally used in conjunction with invisible ultraviolet radiation.

Color names recognized by human observers apply to hues over a spectral interval; the range and major wavelengths are given in Table 9.

Table 9. Perception of spectral color

Color name	Range (nm)	Major wavelength (nm)
Violet	400–440	439
Blue	440–500	472
Green	500–570	512
Yellow	570–590	577
Orange	590–610	598
Red	610–700	630

66 Ultraviolet-Visible Radiation

For practical purposes, the range of wavelengths of ultraviolet-visible radiation relevant for analytical spectroscopy and effects on food extends from 200 to 780 nm.

2.2 Interactions Between Ultraviolet-Visible Radiation and Matter

2.2.1 Absorption and Emission of Light

The same physical rules govern the mechanisms of molecular absorptions in both the ultraviolet and visible regions of the spectrum, and therefore it is justified to treat them jointly. The energy absorption in these spectral ranges results in the elevation of valence electrons from ground-state orbitals to the higher energy orbitals of an excited state. The molecules in these electronic excited states are more reactive than the ones in the ground state and are more predisposed to undergo chemical modifications. The science of photochemistry is concerned with the chemical and physical effects associated with electronic excitation of chemical compounds.

The first law of photochemistry requires that the radiant energy must be absorbed by the target molecule in order to induce a photochemical reaction. Since the process of absorption is quantized, the energies which can be absorbed by a particular compound correspond exclusively to the wavelengths of the bands in its absorption spectrum. The absorption wavelengths are determined by the type and arrangement of component atoms in the molecule.

The second fundamental law of photochemistry, which is a result of the quantum theory, states that only one molecule is excited for each quantum of absorbed radiation. This implies that the efficiency of a photochemical process can be defined by its quantum yield (ϕ), which is the ratio between the number of molecules undergoing a particular change and the number of quanta absorbed. In most photochemical reactions the quantum yield will range from zero to unity. However, in a chain reaction, although the absorption of one photon can initiate one reaction at most, the subsequent propagation of this reaction in the dark may produce many more molecules of product. In such a case, the quantum yield for the product formation may reach several powers of ten.

To quantify the chemical processes it is convenient to express the molar excitation energy of light, E, *viz.*, an Einstein, in terms of wavelength, λ:

$$E = N\frac{hc}{\lambda}$$

where N is Avogadro's constant (6.022×10^{23}). A useful form of this expression is:

Absorption and Emission of Light

$$E = \frac{28369}{\lambda(nm)} \ kcal/mol.$$

The absorption of radiation by a homogeneous sample of material placed in a beam of radiation was developed by Pierre Bouguer and Johann Lambert in 1760 (Bouguer 1760; Lambert 1760). They demonstrated that the proportion of radiation absorbed by a material depends only on the number of incident photons and is independent of the intensity of radiation. The decrease in light intensity (dI) on passage of a wave through a distance db of a homogeneous layer of material is proportional to the incident intensity:

$$-dI/db = \alpha I.$$

The variables can be rearranged:

$$-dI/I = \alpha \, db,$$

where α is the constant of proportionality. This constant is characteristic for the material at a specific wavelength of radiation. The negative sign reflects the decrease, or absorption, of radiation. Upon integration we obtain:

$$\int_{I_0}^{I} dI/I = -\alpha \int_{0}^{b} db$$

$$\ln I/I_0 = -\alpha b$$
$$I = I_0 e^{-\alpha b}.$$

The Bouguer–Lambert law states that there is an exponential decrease in radiation intensity with the length of passage of radiation through a homogeneous material.

In 1852, August Beer (Beer 1852) found that the transmission of radiation by a homogeneous material depends only upon the number of molecules through which the radiation travels, that is, the absorption is directly proportional to the number of absorbing molecules. The number of molecules in a volume is given by the concentration, c. Beer's law is:

$$dI/I = -\beta dc.$$

The Bouguer–Lambert law may be combined with Beer's law to provide the law of absorption of radiation in a cross-section of homogeneous concentration:

$$dI/I = -\kappa \, c \, db,$$

where κ is the molar constant of proportionality. It follows that upon integrating over an entire path length, b, or concentration, c:

$$\int_{I_0}^{I} dI/I = -\kappa c \int_{0}^{b} db = -\kappa b \int_{0}^{c} dc$$

$$\ln (I/I_0) = -\kappa cb \quad \text{or}$$
$$\ln (I_0/I) = \kappa cb$$
$$I = I_0 e^{-cb}$$

Customarily, the Beer–Bouguer–Lambert law is expressed in common logarithms:

$$\log(I_0/I) = \kappa cb/\ln 10 = \varepsilon cb.$$

In these equations I and I_0 are intensities of transmitted and incident light, c is the concentration of the absorber (mole/l), b is the depth of absorber through which the light has passed or optical path (cm), and ε is a constant of proportionality, specific for the absorber, and known as the "extinction" coefficient $(M^{-1} cm^{-1})$. The extinction coefficient has a characteristic and constant value at each absorption wavelength for a certain chemical structure. In practice, in the advent of molecular associations, the extinction coefficient may also be affected by the concentration of the absorber and the nature of the medium. The product εcd is also named the optical density (OD) or absorbance (reciprocal of transmittance). The linear additive property of optical density with concentration is the basis for its use in chemical analysis.

An alternative measure of absorption intensity, which can be related more readily to theoretical principles, but is unfortunately less suitable for practical applications, is the oscillator strength, f, given by the equation:

$$f = 4.315 \times 10^{-9} \int \varepsilon \cdot dv$$

The major difference between oscillator strength and extinction coefficient is that the former is a measure of the integrated intensity of absorption over a whole band, whereas ε is a measure of the intensity of absorption for a single wavelength.

This treatment for light absorption applies only to the simplest of materials: homogeneous. Most foods are far from homogeneous. Light incident upon a heterogeneous material is scattered, reflected, absorbed, and transmitted. No rigorous theory of multiple scattering exists but phenomenologic treatments, which include absorption and scattering, have been advanced. Scattering increases with decreasing wavelength and with decreasing size of obstructing particles; for particles commensurate in size with the wavelength of light, it depends upon the size, shape, and refraction. The immediate consequence of scattering is to greatly intensify the absorption by up to several hundredfold.

Each chemical molecule has an individual absorption spectrum which is determined by the types and arrangements of its constituent atoms. The configuration of atoms and chemical bonds responsible for light absorption is known as the chromophore. The process of electromagnetic energy absorption in the ultraviolet-visible range is referred to as excitation. Several physical processes might be involved in the photochemical excitation of a molecule and these are represented in Fig. 8.

In the absence of an electronic excitation a molecule exists in the electronic ground state ^0S. Any electronic excited state of molecules or atoms contains many vibrational levels which, in turn, accommodate levels of rotational energy. The energy separation between these vibrational levels is a tenth to a hundredth of that between two successive electronic states. Following the absorption of a quantum of energy, a photon, the molecule is raised to an excited state with higher energy. The absorption of photons in the near or medium infrared regions results in excitation jumps between vibrational levels, and in the far infrared between rotational levels. Since the energy separations between rotational levels are very narrow, these transitions are practically continuous. At any temperature above absolute zero a molecule jiggles between vibrational and rotational levels of the electronic ground state. So, rather than a unique value for an electronic level, there is a range of values determined by the variations created by the molecular vibrations. The absorption of a photon in the ultraviolet-visible range causes excitation from a lower to a higher level of electronic energy, like from the ground singlet state, ^0S, to the first excited singlet state, ^1S. Since each electronic energy level within a molecule is associated with several, very densely packed vibrational and rotational levels, a transition

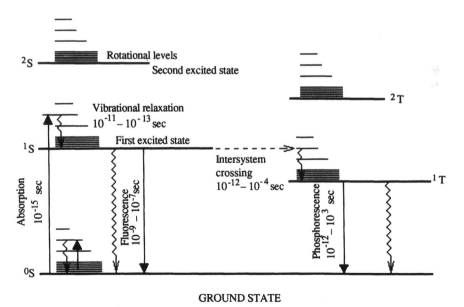

Fig. 8. Electronic transitions following photochemical excitation (Joblonski diagram)

between electronic states is inevitably accompanied by some vibrational and rotational transitions within the newly reached electronic state.

Since the absorption of a quantum of radiation in the ultraviolet-visible range leads to the electronic excitation of the absorber, the electronic distribution of the absorbing molecule is changed as a result of this excitation. Two electrons are assigned to each molecular orbital and their spins are paired. In general, a stable molecule in its ground state has all electrons spin-paired. The initial electronic excitation which takes place from the ground state to an excited singlet state occurs without any change in the orbital spin, i.e., the molecule, although in the excited state, still preserves the spin-paired or "singlet" multiplicity. Very few chemical transformations happen directly from excited singlet states because of their short lifetime (usually less than 100 ns). The molecule in the excited singlet state may decay to the original ground state by converting the excess energy into heat or by spontaneous photon emission. This emission of light from an excited singlet state is called fluorescence. Since some energy is always lost by vibrational relaxations within the excited singlet state, the emission of fluorescence occurs at a longer wavelength than the exciting absorption.

Alternatively, the singlet excited state may undergo a process named intersystem crossing, which involves a spin flip of one of the electrons in the highest occupied molecular orbital. As a result of the spin flip, the system becomes spin-parallel. The quantum theory predicts that such a molecule can exist in three forms of slightly differing, but normally indistinguishable, energy and the molecule is said to exist in a "triplet" state. Since transitions between states of different multiplicities, such as singlet and triplet, are "forbidden" because they involve a spin inversion, the reversal from the excited triplet state to the ground singlet state is slow and the molecule in the triplet state is long-lived (microseconds to seconds). A molecule in the triplet state faces several decay paths. The spontaneous radiation emission from the triplet state is known as phosphorescence. A triplet state always has a lower energy than the corresponding singlet state because of the repulsive nature of the spin–spin interaction between electrons of the same spin and, therefore, the phosphorescence emission takes place at wavelengths that are longer even than fluorescence wavelengths. In general, compounds that fluoresce or phosphoresce are those that contain heteroatoms and multiple conjugated double bonds (aromatic rings). Alternative to or simultaneously with phosphorescence decay, the triplet molecule may react chemically. Because of its long lifetime the triplet-state molecule is the most important mediator in the initiation of photochemical reactions. Having two unpaired electrons, molecules in their triplet states behave chemically as biradicals, and are very susceptible to collisions with other molecules in the medium, including oxygen. Triplet-state molecules readily abstract hydrogen atoms or add to double bonds inter- or intramolecularly.

Absorption and Emission of Light

Finally, a molecule in the triplet excited state may also dispose of its energy by transferring it to an adjoining suitable molecule. The concept of energy transfer can be represented in a general form as follows:

$$^3D^* + {}^1A \rightarrow {}^1D + {}^3A^*,$$

where D and A are the donor and acceptor molecules respectively, in the triplet and singlet states. The donor is the electronically excited triplet molecule and the acceptor is a different molecule present in the medium in the ground state. Excitation of a ground-state molecule (^1A) by energy transfer from another excited species ($^3D^*$) is called sensitization. Viewed from the opposite facet of deactivation of an excited sensitizer, the same process is known as quenching. The energy transfer process plays an important and unique role in light-induced chemistry since, by photosensitization, a molecule can reach an excited state without directly absorbing radiation.

In simple terms of molecular orbitals the change of electronic distribution, as a result of absorption of radiation, is visualized as an alteration in the occupation pattern of a set of orbitals. A combination of orbitals of two atoms, along the internuclear axis which creates the chemical bond, results in the formation of two molecular orbitals: a bonding orbital (either σ or π) and a corresponding antibonding orbital (σ^* or π^*). In addition, if nonbonding electrons are present, as in heteroatoms (oxygen, nitrogen), they occupy a nonbonding orbital "n" which, by its nature, does not possess an antibonding counterpart. The electronic excitation promotes an electron from one molecular orbital to another unoccupied molecular orbital of higher energy, for example, from n or σ to σ^* (n, $\sigma \rightarrow \sigma^*$), or from n or π to π^* (n, $\pi \rightarrow \pi^*$) orbitals. In general, excitation of an electron to the σ^* orbital requires an amount of energy corresponding to wavelengths below 200 nm and these are not accessible from conventional illumination sources. The other excitation processes ($\pi \rightarrow \pi^*$ and n $\rightarrow \pi^*$) are responsible for the bulk of photochemical reactions, since the wavelengths associated with these two transitions are located in the accessible regions of ultraviolet and visible parts of the electromagnetic spectrum.

The $\pi \rightarrow \pi^*$ transition is the longest wavelength, lowest energy transition and is therefore the major electronic transition in olefins exposed to light. Although the antibonding state (π^*) may be looked upon as a biradical state, a bonding electron is still involved in the $\pi \rightarrow \pi^*$ transition and the excess spin density is distributed throughout the π electron system of the molecule, thus diminishing any specific free radical character.

The transition n $\rightarrow \pi^*$ occurs in molecules containing heteroatoms, such as oxygen, sulfur, and nitrogen and arises from the excitation of the electron in the nonbonding orbital of the heteroatom to the π^* antibonding orbital. The carbonyl group is the typical chromophore predisposed to this transition. The alterations in molecular conformation resulting from n $\rightarrow \pi^*$

transitions lead to substantial differences in reactivity as compared to the ground state. The configuration and behavior are analogous to a biradical on the heteroatom and are therefore highly reactive in hydrogen abstraction reactions with vicinal molecules or intramolecularly, if the configuration allows. In practical terms, this means that a molecule that is less saturated and which contains heteroatoms needs a lower excitation energy and absorbs at a longer wavelength than does a more saturated molecule.

2.2.2 Polarized Light

Visible light, being an electromagnetic radiation, behaves as if it had an electric component and a magnetic component which act as though they were at right angles to each other. The directions of the electric and magnetic components are purely at random. If, however, the electric and magnetic vectors of all the photons in a beam of light are in the same plane, as is the case for light emerging after passage through certain kinds of mineral crystals, the light is called plane polarized. Circularly polarized light, in which the electric vector rotates and traces out either a left-handed or right-handed helix, can also be generated. A beam of left-handed circularly polarized light together with a comparable beam polarized in the right-handed direction is equivalent to a beam of plane-polarized light.

A transparent solution of a compound with a plane of symmetry does not affect polarized light. If, however, polarized light strikes an asymmetric molecule, it emerges from the solution with the plane of polarization rotated. The rotation of polarized light does not involve light absorption and electronic excitation. This phenomenon plays a fundamental role in analysis of sugars and other optically active compounds.

2.3 Photooxidation

A photosensitization process of particular interest in food chemistry is photoxidation. Since the early stages of photochemistry, chemists have been aware that light is one of the means of initiating oxidations of organic materials.

In ground-state oxygen the two electrons of highest energy occupy two degenerate π^* molecular orbitals (2py and 2pz). Following Hund's rule these two electrons have their angular momentum opposed but the spins are parallel. Such an electronic configuration is designated in spectroscopic language as $^3\Sigma_g^-$ which indicates that ground-state molecular oxygen, O_2, exists as a triplet molecule. This triplet character is responsible for the paramagnetism and diradical-like behavior of $^3O_2$. More importantly, this

Photooxidation 73

triplet electronic configuration only permits reactions involving one-electron steps. Thus, despite the exothermicity of oxidation reactions, a spin barrier prevents $^3O_2$ from reacting indiscriminately with the plethora of singlet ground-state compounds surrounding it.

A higher energy configuration, in which these two electrons are located in the same π^* (2py) orbital, corresponds to the lower excited state of the oxygen molecule, which is a singlet state because the electrons are paired and angular momentum is the same. This state of the oxygen molecule is denoted as $^1\Delta_g$ and lies 22.5 kcal/mol above the ground state (corresponding to a photon energy of approximately 1 eV = 1270 nm). The next excited state of the oxygen molecule is 37.5 kcal/mol above the ground state and has each of the π^* (2p) orbitals half full. It is denoted by the symbol $^1\Sigma_g^+$ and it differs from the $^3\Sigma_g^-$ state in that the last two electrons have antiparallel spins (Table 10).

In the gas phase the lifetimes of $^1\Delta_g$ and $^1\Sigma_g^+$ oxygen are 45 min and 7 s respectively. However, in water solution these lifetimes are dramatically reduced through collisional deactivation to approximately 10^{-3} (somewhat longer in lipids and other nonpolar media) and 10^{-11} seconds respectively (Foote 1976; Frimer 1985). Because the reactions of concern to food chemistry are generally carried out in solution, it is the longer lived O_2 ($^1\Delta_g$) that is involved as the active species.

The most likely source of $^1O_2$ in foods is photosensitization. Molecular oxygen in the gas phase does not absorb ultraviolet or visible radiation, but quenches, with a diffusion-controlled rate, the excited triplet state of many organic molecules. The energy level of $^1O_2$ of 22.5 kcal/mol is lower than most excited organic molecules which absorb in the visible range (carotenes are a notable exception) and therefore the quenching is very efficient. As a result of this energy transfer, the molecular oxygen is elevated to the much more reactive excited state:

$$^3\text{Organic sensitizer} + {}^3O_2 \rightarrow {}^1\text{organic sensitizer} + {}^1O_2.$$

Since both the sensitizer and ground-state oxygen are triplets, their interaction does not require a change in spin direction and it is very efficient.

Table 10. The three lowest electronic states of molecular oxygen

State of O_2 molecule	Symbol	Relative energy (kcal/mol^{-1})	Orbital occupancy	
Second excited	$^1\Sigma_g^+$	37.5	↑	↓
First excited	$^1\Delta_g$	22.5	↑↓	
Ground	$^3\Sigma_g^-$	0	↑	↑

The chemical reactions of singlet oxygen with organic molecules differ from those of ground-state oxygen. Singlet oxygen does not face the spin restriction of the diradical ground state and reacts more rapidly, albeit somewhat more selectively, with many electron-rich compounds to give oxidized species. The photosensitized oxidation proceeding by a singlet oxygen mechanism is known as type II photoxidation.

There are four types of singlet oxygen reactions which may be relevant in food systems:

1. The "ene" type reaction occurs with olefins having two or more alkyl substituents and results in the formation of allyl peroxides in which the double bond has shifted to a position adjacent to the original double bond. The most common example of this reaction in foods is the formation of lipid peroxides when $^1O_2$ reacts with unsaturated fatty acids or cholesterol (Fig. 9). Free radical oxidation of the same lipids usually yields a more complex mixture of hydroperoxides, and products with unshifted double bonds are formed (Frankel 1980, 1985).
2. The [4 + 2] Diels-Alder type addition to dienes and heterocyclics frequently results in the formation of endoperoxides of varying stability (Fig. 10).
3. Dioxetane formation can occur when electron-rich olefins and enamines react with $^1O_2$. These products are frequently unstable and undergo facile cleavage resulting in the formation of carbonyl compounds (Fig. 11).
4. Compounds with heteroatoms, such as sulfur or nitrogen, can be oxidized by $^1O_2$. Sulfides such as methionine are oxidized to sulfoxides (Fig. 12). Nitrogen-containing compounds can undergo N-oxide formation.

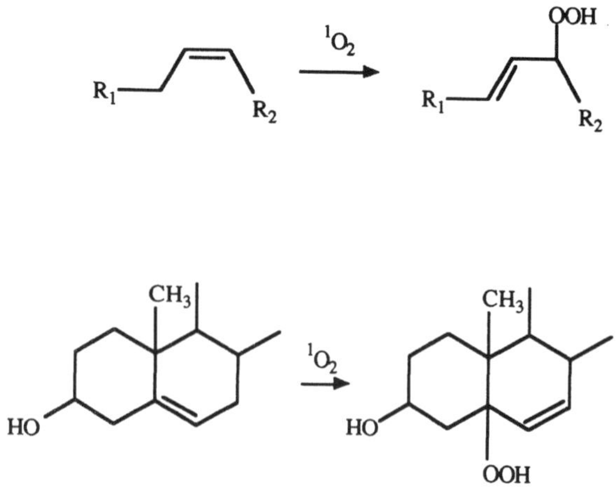

Fig. 9. "Ene" type oxidation

Photooxidation

In a genuine photosensitized reaction, like the generation of singlet oxygen, the triplet sensitizer ultimately returns unchanged to the ground state and can absorb another photon, thus acting in a somewhat catalytic fashion. In other cases, however, the light-absorbing species can be consumed in the reaction. This happens when the triplet sensitizer reacts irreversibly with molecules ("substrate") in its vicinity to convert them to new chemical species. In general, a sensitizer molecule in its triplet state can both abstract or donate electrons or hydrogen atoms to other molecules more readily than in the ground state. The efficiencies of these reactions depend on the chemical structures of the sensitizer and the substrate, as well as on the reaction conditions. Thus, the reduction of the triplet sensitizer by a substrate proceeds as follows:

$$^3\text{Sens} + \text{RH} \rightarrow \text{Sens} \cdot^- + \text{RH} \cdot^+$$
$$^3\text{Sens} + \text{RH} \rightarrow \text{SensH} \cdot + \text{R} \cdot.$$

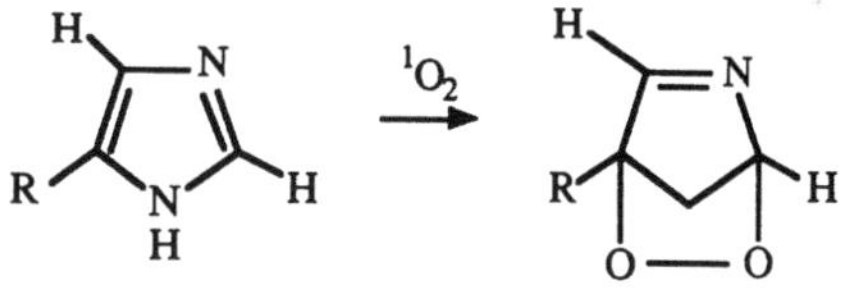

Fig. 10. The addition of $^1O_2$ to imidazole

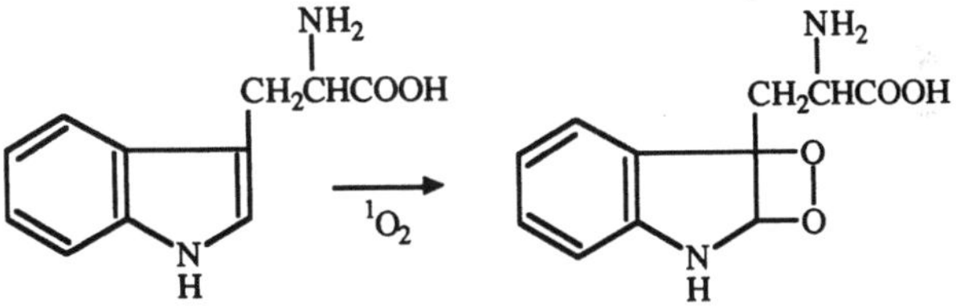

Fig. 11. The addition of $^1O_2$ to the imidazole ring

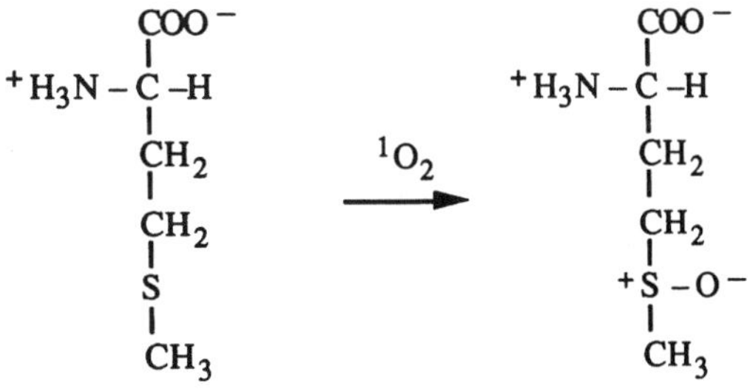

Fig. 12. Addition of $^1O_2$ to methionine

These free radical products are very reactive chemically. In many cases, the resultant substrate radicals react with ground-state oxygen to give oxidized products of various types; often these are peroxides, which can react further to initiate free radical chain-type autoxidation processes:

$$R \cdot + {}^3O_2 \rightarrow ROO \cdot$$
$$ROO \cdot + RH \rightarrow ROOH + R \cdot$$

The photosensitized oxidations proceeding by a free radical mechanism are known as type I. The final oxidation products of type I oxidations may or may not be different from those produced by singlet oxygen; the sensitizer, substrate, and reaction conditions are crucial.

Occasionally, the ground-state sensitizer can be regenerated by the reaction of the semireduced from with ground-state oxygen $({}^3O_2)$ to give a ground-state sensitizer molecule and the superoxide radical anion, O_2^- (or its conjugate acid, $HO_2 \cdot$):

$$Sens \cdot^- + {}^3O_2 \rightarrow Sens + O_2^-$$
$$SensH \cdot + {}^3O_2 \rightarrow Sens + HO_2 \cdot$$

Superoxide radical anion (O_2^-) is not very reactive and readily disproportionates to H_2O_2 in water. Because of its lack of reactivity, O_2^- cannot be directly responsible for the initiation of the oxidation chain. However, there is a possibility for O_2^- conversion to a very reactive species, such as a hydroxyl radical, as the result of an O_2^--driven Fenton reaction; the transition ion catalyst, M^{n+}, can well be provided by heme proteins:

$$2O_2^- + 2H_2O \rightarrow H_2O_2 + 2OH^- + O_2$$
$$M^{n+} + O_2^- \rightarrow M^{(n-1)+} + O_2$$
$$M^{(n-1)+} + H_2O_2 \rightarrow M^{n+} + OH \cdot + OH^-$$

The interaction between an excited sensitizer and a molecule of substrate is favored at high substrate and low oxygen concentrations, since oxygen effectively competes with the substrate for interaction with the triplet sensitizer. Substrates most efficiently photodegraded by direct interaction with a sensitizer include those that are readily oxidized, such as amines or phenols, or are readily reduced, such as quinones. An example of such a reaction, which occurs in milk exposed to light, is the riboflavin-sensitized photodegradation of the amino acid methionine. On illumination, triplet flavin is generated and abstracts one electron from the sulfur atom of the amino acid; the resulting methionine radical undergoes deamination and decarboxylation to give methional (β-methylmercaptopropionaldehyde). The reduced flavin radical reacts with oxygen and regenerates ground-state flavin and hydrogen peroxide.

The general end-result of a photochemical process depends on several factors:

Illumination Sources 77

1. The nature of the radiation absorber (its photochemical stability and the nature and reactivity of the chemical species generated by its decomposition; its ability to be photoreduced by intermolecular reactions, as compared to the opportunity for energy transfer to generate singlet oxygen).
2. The nature of the substrates and their redox potentials.
3. The relative concentrations of sensitizer and substrate.
4. The nature of the medium and the freedom of molecular diffusion.

It is important to point out that most photochemical processes have only one initial "light" step followed by one or more "dark" reactions.

2.4 Illumination Sources and Units of Measurement

2.4.1 Illumination Sources

There are two effective ways of producing artificial light for general illumination: incandescent and fluorescent lamps. In incandescent lamps, the radiant source is a hot filament of tungsten in a sealed tube, which can now be combined with a halogen to improve its efficiency and durability. The incandescent filament has a temperature of approximately 2850 K, so that its radiation is shifted to the far-red end of the spectrum and gives the light its "warm" appearance compared with standard daylight at 6500 K. Indeed, about 90% of the total emission of an incandescent lamp lies in the infrared. Incandescent lamps are inefficient, generating more heat than light, but the light is of good quality, illuminating the colors of objects realistically.

Unlike sun or incandescent lamps, fluorescent lamps generate light by a nonthermal mechanism. Fluorescent lamps are very efficient, providing up to 200 lm/W compared with 12 lm/W from an incandescent lamp, but their design and operation are more complicated and the quality of light is variable. In fluorescent lamps an electric arc discharge is established through the mercury vapor, present in the tube at low pressure (0.008 torr), when the proper voltage is applied to the electrodes at the ends of the tube. The energy of the discharge elevates the mercury atoms to an excited state and they revert to the ground state by emiting mostly ultraviolet radiation, primarily at 253.7 nm and, to a lesser extent, at 297, 303 and 313 nm. Indeed, germicidal lamps, which have transparent and colorless quartz tubes containing mercury vapor at low pressure, emit 95% of their energy in the 253.7 nm mercury line. In the common types of fluorescent lamps used for illumination, this ultraviolet radiation is converted into visible light by a phosphor coating on the inside of the glass tube. The spectral distribution of the optical radiation of fluorescent lamps depends upon the blend of the fluorescent chemicals in this phosphor coating (usually appearing, when not lit, as a matt white, translucent coating) on the inside of the tube. The different types of fluorescent lamps, "warm white", "daylight", "cool light", and "soft white", have common spectral characteristics. These lamps emit optical radiation over a broad wavelength range, in a combination of continuous and line spectra, from within the ultraviolet, through the visible and up to 800 nm into the infrared. The tube of the fluorescent lamps is made of glass, since the absorption of this material is very high at wavelengths shorter than 320 nm and, at the same time, is very transparent in the visible range. Thus, although almost all fluorescent lamps emit some shortwave ultraviolet radiation in the 300 nm wavelength region, primarily in the emission lines of mercury at 297, 303 and 313 nm, this emission is minimized. In contrast, silica glass (quartz), which is used as the envelope for germicidal lamps, is transparent down to 200 nm.

The characteristic spectral emissions for sunlight, incandescent, and fluorescent lamps are shown in Fig. 13.

Unlike the incandescent lamps, fluorescent lamps emit very little infrared radiation. This explains their virtual monopoly for lighting food areas. Indeed, fluorescent lamps are the predominant source of illumination in processing, storage, and display areas for food. Incandescent lamps, which have a very high output in the infrared, cannot be used close to foods that would be damaged by high temperatures.

A special type of fluorescent lamp, commonly referred to as "blacklight" (or Wood's lamp), consists of a high-pressure mercury arc equipped with a

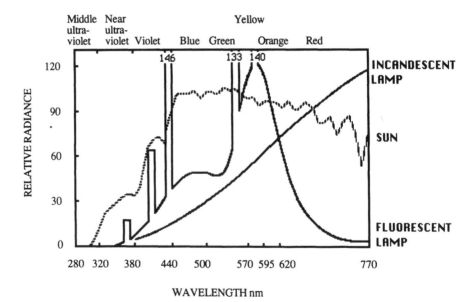

Fig. 13. Emission spectra of common sources of light

Biological Effects and Safety Aspects of Ultraviolet-Visible Radiation

nickel oxide phosphor coating; this has emission primarily in the 350–370 nm range and a negligible emission in the visible.

2.4.2 Units of Measurement

A light source is defined by the spectral distribution and the energy of the emitted radiation. Many different terms and units are used to describe the energy of ultraviolet-visible radiation: ergs, joules, calories, or electron volts (Thimijan and Heins 1983). The conversion factors between these units are as follows:

$$1\,J = 10^7\,erg = 1\,W.s. = 6.25 \times 10^{18}\,eV = 2.38 \times 10^{-4}\,kcal$$

The intensity of radiation is defined as the energy delivered per unit of time and cross section and is most commonly expressed in W/cm^2 or equivalent units $(1\,W/cm^2 = 9.29 \times 10^2\,W/ft^2)$. To make the information relevant for understanding a photochemical process the wavelength region must be specified.

The intensity of a source of visible light can also be measured in candle power. The candela (candle) is 1/60 of the luminous intensity of $1\,cm^2$ of a black body radiator at the freezing temperature of platinum (2046 K). More recently, a candela was defined as the luminous intensity of a source emitting monochromatic radiation with a frequency of $540 \times 10^{12}\,Hz$ and a radiant intensity of 1/683 W per unit solid angle (steradian). This corresponds to about 54.5 candle power to 1 W at 556 nm, which is the wavelength of maximum visibility. The total energy of visible energy emitted by a source per unit time is the total luminous flux. The unit of flux, the lumen (lm), is the flux emitted in a steradian by a point source of 1 cd. Thus 1 cd intensity emits $4\pi\,lm$. The density of the luminous flux on a surface is called illumination. It is the quotient of the flux divided by the area of the surface when the latter is uniformly illuminated. The foot-candle (ft-c) corresponds to the illumination of 1 lm incident per square foot; alternatively, 1 lx equals $1\,lm/m^2$. The conversion factors between the various units of light intensity are as follows:

$$1\,W = 5.8 \times 10^5\,ft\text{-}c\ (at\ 556\,nm) = 6.28 \times 10^6\,lx\ (at\ 556\,nm).$$

2.5 Biological Effects and Safety Aspects of Ultraviolet-Visible Radiation

As early as 1877 a bactericidal action of ultraviolet light applicable equally to moist and dry materials was recorded. It was first thought that free oxygen was necessary but the idea was disproved soon afterwards. In the period 1909–1915 many patents were taken out for photosterilization of water, wine, and meat, but few of them had much permanent technological success. New ground was broken in 1928 in understanding the mode of bactericidal action when it was realized that the action spectrum (relative spectral effectiveness of radiation in eliciting a particular biological effect) agreed rather well with the absorption spectrum of nucleic acids. The peak absorption by purine and pyrimidine residues in nucleic acids occurs at 260–265 nm and it is this wavelength range that accomplishes most damage. The probability that DNA is the significant absorbing substance in the damaging effect of irradiation has led to in vitro studies on the photochemistry of nucleic acids, nucleotides, and purine or pyrimidine bases. The sugar moieties are practically transparent in the region 230–300 nm. The photosensitivity of pyrimidines is much greater than that of purines, the quantum yields being in the order of 10^{-3} and 10^{-4} respectively. Although energy transfer is not excluded, it is probable that pyrimidine absorption is the major cause of DNA photochemical modification, and therefore underlies the molecular mechanism of biological damage. Some cellular and extracellular physiological changes are attributable to the proteins and their constituent amino acids. To a first approximation the absorption spectrum of proteins is closely allied to the absorption spectrum of specific amino acid residues. The amino acids which contain aromatic residues absorb the most in the ultraviolet region, shorter than 300 nm: phenyalanine, tryptophan, and tyrosine.

Because of man's long-term familiarity with visible light, it is generally regarded as being safe or innocuous. However, this long-term familiarity does not render all types of "light" harmless. Some light sources also produce ultraviolet radiation of potentially adverse effects. For photobiological purposes, the portion of the ultraviolet spectrum that can be transmitted through air is divided into three regions of increasing noxiousness: UV-A, the least energetic portion between 320–400 nm; UV-B between 280–320 nm; and UV-C, the most energetic portion, of wavelengths shorter than 280 nm. The principal recognized hazard of the exposure to ultraviolet radiation was the potential for injury to eyes (Sliney and Freasier 1973) and skin (Pierce et al. 1986) from ultraviolet radiation shorter than 320 nm. Exposure of the eyes to ultraviolet radiation from unshielded lamps can have, in the short term, delayed and very painful effects due to inflammation of the cornea, and, in the long term, can cause cataracts. In extreme cases vision can be irreversibly impaired by ultraviolet irradiation if goggles are not worn.

The skin protects the internal organs and structures from ultraviolet radiation. When ultraviolet radiation enters the skin and blood a host of interactions with biomolecules and their aggregates, such as membranes, enzymes, DNA, and RNA, are initiated, and both physiological and pathological events might be evoked in the skin. The effects can be divided into two general types: acute and chronic. Acute effects include sunburn and

The Color of Foods 81

chronic effects include premature aging of skin and development of certain forms of skin cancer. In addition, the skin is a major site of immunological activity, and ultraviolet radiation is capable of affecting the immune system via its effects on skin. Because of these hazards the exposure of humans to germicidal lamps must be totally avoided.

2.6 Effects of Ultraviolet-Visible Radiation on Foods

2.6.1 The Color of Foods

A primary impact of visible radiation on food is the creation of color. The light initiates biochemical and physiological processes in the complex of the eye–optic nerve–cerebral cortex and these are interpreted in terms of color perception. Thus, the color is the total visual experience resulting from this biological stimulus by certain intensities and wavelengths of light. Perception of food color is a function of the degree of light absorption, reflection, and scattering. The food color is determined not only by the chemical structure of the pigments present but also by their physical state, relative to the nonpigmented constituents. For instance, the color of green vegetables changes from dull to bright green upon blanching. This is not, for the most part, caused by a change in composition of the chromophore-chlorophyll pigments, but is due to a change in their physical state within the chloroplasts, to the expulsion of air, and to an increase in translucency of intercellular components as well as that of surrounding tissue. The perception of carotene color in marine foods depends on the amount and disposition of colorless lipid material (Little 1976). Occasionally, food color is perceived in the complete absence of colorants as, for instance, that elicited by scattering, e.g., the opaque whiteness of milk, the transparency of egg white compared with its whiteness when whipped, or the translucency or whiteness of sugar crystals.

There are three fundamental ways by which colors may be described quantitatively:

1. Systems based on the measurements of sensation in color vision. They are dependent on three basically psychological parameters: hue, saturation, and lightness. Hue is the perception of color resulting from differences in absorption of incident energy at various wavelengths, such as green, blue, yellow, or red. Saturation (chroma or purity) describes the reflection at a given wavelength. The lightness (or value) describes the relation between total reflected and absorbed light, with no regard to specific wavelength. If the light is reflected from a surface evenly at all

angles, the impression of a product with a "flat" or "diffuse" appearance is created. If the reflection is stronger at a specific angle a glossy appearance results.

2. Systems based on the stimuli generated in the human eye by visible light of various wavelengths and intensities. The most widely adopted system of such a color description is the CIE (Commission International de l'Eclairage) system. It is based on the fact that light reflected from any colored surface can be visually matched by an additive mixture of red, green, and blue light in suitable proportions. The basis of the CIE system is three parameters based on three primary colors. The description of colors as color stimuli for the human eye is therefore three-dimensional.

3. The expression of color in accurate spectroscopical terms of wavelengths absorbed or reflected.

Foods may possess a natural color or may be colored artificially during processing by addition of approved synthetic dyes, lakes, pigments, or natural extracts. The natural vegetable pigments belong to one of four chemical classes:

1. Anthocyanines are orange, red, and blue glycosides of substituted flavylium salts. A combination of anthocyanines, typically four to six, are responsible for the color of grapes, strawberries, blueberries, cranberries, apples, roses, red cabbage, avocados, and eggplants.

2. Betalains are the red compounds found only in the family Centrospermae, which includes red beets and cactus fruit. The structural features include dihydroindole and dihydropyridine rings and a polymethylene cyanine chromophore.

3. Carotenes are isoprenoid polyolefins which are yellow, orange, or red and are found in virtually all green plants and in some animal tissues: green leaves, red tomatoes, paprika, carrots, corn kernels, red salmon, and butter.

4. Chlorophylls are the green porphyrin compounds synthesized in huge amounts in nature as the pigment necessary for photosynthesis. The metal porphyrin ring is common to the green color of plants and to the red heme found in meat. The pigments associated with the bright red color on the surface of uncooked meat are oxymyoglobin (MbO_2) and oxyhemoglobin (HbO_2).

A natural food product is subject to a number of variables which affect the natural coloring components. Therefore, the color of a food becomes an important criterion of quality and plays a role in the selection and development of vegetable and animal foods for consumers. Deliberate efforts are made to select tomatoes for redness, oranges with a high carotene content,

Beneficial Effects of Light in Production of Foods 83

poultry meat for lightness, etc. Sensory panels have demonstrated that the anticipated taste is based on the color of the product. People want to eat familiar-looking food and color is one of factors that the consumer is conditioned for. A green tomato is perfect pickled but is unfit for a fresh salad, butter is preferred yellow than white, and a clear cola drink would not match the preconceptions of a brown one.

Color can be an excellent indicator of maturity, quality, and wholesomeness but, on the other hand, can suggest contamination or deterioration. A most frequently encountered spoilage of vegetable food is brown discoloration, appearing in senescent and injured plant material. These browning products vary in hue from yellowish to reddish brown through to dark brown. By and large, the reaction is due to melanins formed as a consequence of the action of enzymes known as phenolases, which catalyze the oxidation of phenolic substances. The common browning reaction is a response to a variety of stresses. Causative agents are mechanical damage due to bruising, excessive pressure, infection due to molds, and bacteria. In all these cases, cell disruption allows phenolic substrates to become accessible to the action of phenolases.

Color evaluation is a powerful analytical tool in following the success of processing conditions, not only for foods that are processed to retain their natural color but also for those that are refined to whiteness or are treated to develop color, such as by roasting. Length of shelf-life and determination of optimum conditions for storage can be determined by following color changes, and other quality factors, with time.

The overriding problem with color is that of obtaining an objective, precise and reproducible procedure for measuring quantitatively the color differences which can be perceived and the attributes: reflection (matt, opaque, or glossy), transparency, or turbidity. Modern absorption and reflectance spectrophotometers have replaced colorimeters and color comparators. A spectrophotometric curve, that is, the plot of intensity versus wavelength, is certainly the most unambiguous specification of color that can be obtained.

2.6.2 Beneficial Effects of Light in Production of Foods

In the course of time research efforts have been invested, on the one hand, in utilizing artificial photochemical reactions for the solution of practical problems in food processing and, on the other hand, in avoiding detrimental effects of light on foods.

No food of animal or vegetable origin can be produced without light. The photochemical reaction cardinal to food production, photosynthesis, is beyond the scope of this book which covers artificial processes. However, a few related reactions performed on harvested commodities ought to be

84 Ultraviolet-Visible Radiation

mentioned. Thus, exposure to artificial light has been used in attempts to improve the quality of apples. One of the characteristics that determines the marketability of apples is fruit color. The postharvest exposure to ultraviolet and white light has a synergistic effect on the enhancement of anthocyanine production, thus inducing the red coloration, but does not affect ripening; of mature green apples (Bishop and Klein 1975; Arakawa et al. 1985; Sakes et al. 1990). A similar stimulation of anthocyanine synthesis has been reported in milo sorghum (Mohr and Drumm-Herrel 1981) and in several plant seedlings with blue plus ultraviolet light (Mohr and Drumm-Herrel 1983), and in broom sorghum with red plus ultraviolet light (Hashimoto and Yatsuhashi 1984).

Compared to the overriding importance of natural light in photosynthesis and production of food, artificial photochemistry has only very limited applications. The ultraviolet radiation of wavelengths shorter than 300 nm, such as that emitted by germicidal lamps (253.7 nm), is lethal to microorganisms. Critical biocellular components, including proteins and nucleic acids, absorb this wavelength very strongly and are subsequently prone to chemical modifications which are accompanied by disastrous biological results (Proctor and Goldblith 1951). These lamps are successfully used to decrease the atmospheric counts of microorganisms in storage areas and thus to disinfect and avoid the contamination of foods. Germicidal lamps are also used for sterilization of inner surfaces of packaging materials, particularly in aseptic packaging lines. However, these low-pressure mercury discharge lamps always emit also a small amount of light of longer wavelengths which may diminish the effectiveness of decontamination treatments. Light of wavelengths between 325 and 550 nm is known to cause photoreactivation of vegetative cells damaged by short ultraviolet and it was suggested that a similar photoreactivation phenomenon might occur on food in contact with reflective surfaces. Indeed, smoothness, reflectivity and geometry of the surfaces of food packaging are important factors in the extent or variability of the lethal effect of ultraviolet radiation. Surfaces containing aluminum in the laminate reflected more light in the 325–550 nm range and showed a lower lethal effect when spores of *Bacillus subtilis* were irradiated. The geometry of the irradiated surface is important for aluminum/polyethylene laminate-lined surfaces only, as more spores were killed on a surface normal to incident 254 nm irradiation than in cartons with reflective angles. Spores attached on the inner sides of this type of carton may have received more reflected light of photoreactivation wavelengths (Stannard et al. 1985). The ultraviolet treatment for disinfection has also been used in combination with hydrogen peroxide (Stannard and Wood 1983; Waites et al. 1988). Ultraviolet irradiation decreased the percentage of *Fusarium* rot of sweet potatoes during storage. The optimum dose was approximately $4 \times 10^4 \, erg/mm^2$. The effect of ultraviolet irradiation on potato nutrients was not significant, except that starch content was

Beneficial Effects of Light in Production of Foods

higher for ultraviolet-irradiated than for nonirradiated roots (Stevens et al. 1990). Ultraviolet radiation has also been tested on control of storage and quality of walla walla onions (Lu et al. 1988). A study was conducted to assess whether the combination of ultraviolet and low-dose gamma radiation could decrease fungal decay incidence in papaya to enhance its market-life. The combined treatment prevented growth of two fungi (*Phytophthora* spp. and *Colletotrichum* spp.) but not that of a third fungus (*Aschochyta* spp.) studied. In most cases, the combined treatment did not improve the market-life of papaya (Moy 1977). Low-pressure mercury-vapor ultraviolet lamps were not effective in decreasing airborne mold spores in the cheese-curing room. Direct irradiation failed to prevent mold growth on the surface of cheese (Smith 1942).

The attempts to improve the bacteriological quality of foods by irradiation with ultraviolet of short wavelengths are, in general, hampered by the shallow depth of penetration and the almost unavoidable chemical modifications of the sensitive food constituents. Ultraviolet radiation of wavelength 253.7 nm was effective in destroying bacteria on the surface of fresh mackerel fish (Huang and Toledo 1982), beef slices (Kaess and Weidemann 1971), and fresh meat (Reagan et al. 1973; Stermer et al. 1987). Ultraviolet radiation was also very effective in reducing bacteria counts on agar plates, but less so on beef samples since only exposed bacteria could be killed. For example, the treatment was more effective on the smooth surface created by the meat fibers parallel to the surface than on steaks cut across the fibers.

Another photochemical reaction with attractive potential from a nutritional point of view is the conversion of provitamin to vitamin D_3 in lipids of animal origin (Fig. 14).

To determine the photochemical conversion of provitamin D_3 in fish meat, dark muscle was irradiated with ultraviolet light of $\lambda_{max} = 305$ nm under various conditions of exposure time and intensity of light. After

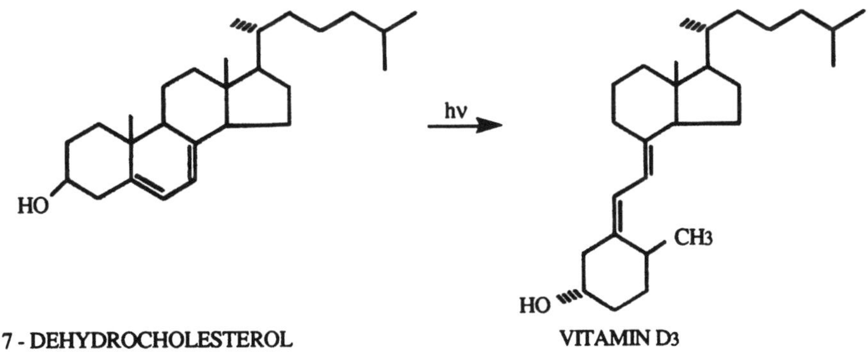

7 - DEHYDROCHOLESTEROL   VITAMIN D3

Fig. 14. The photochemical synthesis of vitamin D

exposure to 20 W light at a distance of 50 cm, the content of vitamin D_3 increased and that of provitamin D_3 decreased. The conversion reached its maximum of 20% after 15 h irradiation (Hidaka et al. 1989).

The possibility to increase the vitamin D content of milk by photochemical conversion was also considered. It is interesting that a product called "irradiated milk" was apparently introduced commercially in the past, as a vitamin D-enriched milk. In Sommer (1938) the process is described as follows:

Irradiated milk is produced by exposing milk in thin, flowing films to ultra-violet radiations from a carbon arc lamp or a quartz-mercury vapor lamp. Several different types and capacities of equipment are available for the commercial irradiation of milk. In each case the conditions required to produce the desired potency have been determined by experimentation, and in commercial operations these conditions are to be duplicated. The essential conditions pertain to the radiant energy output of the lamp, and its distance and location with respect to the milk film, the area of the film, and its thickness as determined by the flow rate insofar as these factors determine the energy to which the milk is exposed. For a given piece of equipment commercial control of the process resolves itself into fixing the flow rate of the milk, automatic recording of the radiant energy, and periodic bioassays of the milk.

Carbon Arc Equipment. The first commercial irradiation of milk with carbon arc lamps was accomplished by flowing the milk over surface coolers and using lamps with reflectors to distribute the radiations over the flowing milk film. Equipment of this type is still in use and gives satisfactory results. The arc lamp is surrounded by the milk film as it flows over the inside surface of a hollow, inverted, truncated cone. The milk enters at the top into a distributing trough, and is collected at the bottom in a trough. In the process of irradiation the shorter wave lengths cause the dissociation of atmospheric oxygen and some of it recombines as ozone. In order to eliminate the ozone and gases from the arc, the air surrounding the lamp is renewed by fan ventilation. One type of equipment has been developed to shield the milk film by interposing a water film. By discharging the liquid from a circular slit of suitable design a continuous, unsupported, cylindrical, flowing film can be created, similar to a large bubble open at opposite ends. The arc lamp in the center is surrounded by the flowing water film, and this in turn is surrounded by the flowing milk film; the atmosphere confined by the water film is renewed by ventilation. This equipment has not come into extensive use, except in the irradiation of yeast.

The spectrum of the radiations from a carbon arc lamp can be varied by impregnating the core of the electrodes with various metals. Special carbons have been selected and are recommended on the basis of the spectrum of their radiations with the aim of obtaining maximum activation with a minimum effect on flavor.

Quartz-Mercury Vapor Lamp Equipment. The spectrum of the mercury vapor arc is suitable for the irradiation of milk. The lamp must be constructed of pure quartz glass since other glasses do not satisfactorily transmit radiations of the shorter wave lengths. With long continued use even the quartz glass loses some of its ability to transmit short wave lengths, but the lamps can be reconditioned by suitable heat treatment. Equipment employing quartz-mercury vapor lamps has been developed and is in commercial use for the irradiation of milk.

Beneficial Effects of Light in Production of Foods

Irradiated Flavor. Over-irradiation of the milk produces a characteristic irradiated flavor which is similar to the flavor produced in milk by exposure to sunlight. The flavor resembles a burnt taste, and apparently originates from the proteins. It becomes intensified when the milk is subsequently heated to high temperatures such as 180 °F or over. The occurrence of this defect is analogous to the occurrence of a cooked flavor in pasteurized milk; both are caused by over-treatment in their respective processes, and in both the over-treatment may be general or localized. In the case of irradiation the over-treatment may result from the use of too much radiant energy in relation to the rate of milk flow (general), or from uneven distribution of the milk film so that certain parts are over-exposed (localized). When the process is carried out under prescribed conditions, no perceptible flavor is developed.

However, the benefit of irradiating milk is outweighed by the development of an unpleasant flavor and by the destruction of photosensitive vitamins (Sattar and deMan 1975), and the irradiation of milk is no longer practiced. Today, the addition of vitamin D to milk is the preferred method of enrichment.

The duality of possible benefits versus negative effects on food wholesomeness plagues any attempt to photoaffect selectively only certain constituents in food. Some contaminants, like chloro-organic pesticides and aflatoxins, absorb and are sensitive to light of longer wavelengths and, indeed, exposure to ultraviolet radiation of contaminated whole milk or melted butter led to the destruction of methoxychlor pesticide. Unfortunately, unpalatable flavors were also developed (Li and Bradley 1969). Likewise, polychlorinated biphenyls in artificially contaminated shrimps were destroyed by exposure to sunlight (Khan et al. 1976). Aflatoxin detoxification by sunlight exposure was attempted in milk casein (Yousef and Marth 1985) and in groundnut (Shantha and Murthy 1981).

A typical singlet oxygen reaction has been suggested for the removal of traces of oxygen from sealed packages of oxygen-sensitive foods as an alternative to vacuum packaging, as well as from microbiological jars for anaerobic plates. A film of polymer, such as ethyl cellulose, cellulose acetate, or rubber, containing a dissolved photosensitizing dye, such as tetraphenyl porphine or erythrosine, and a singlet oxygen acceptor, such as difurfurylidenepentaerythritol, was sealed in the headspace of a transparent package. On illumination of the film with light of the appropriate wavelength excited dye molecules excite oxygen molecules, which have diffused into the polymer, to the singlet state. In turn, these singlet oxygen molecules diffuse to react with acceptor molecules, which are essentially immobile in the matrix, and are thereby consumed. The scavenging process continues as long as the polymer is illuminated and until all the acceptor or headspace oxygen has reacted (Rooney 1981, 1982, 1983; Rooney et al. 1981).

Ultraviolet light has also been employed as a detection tool. The "blacklight" lamp is used occasionally to detect the presence of insects, rodent excreta, fungi, and other contaminants in certain solid, dry foods, by creating a visible fluorescence in the dark (Spikes 1981). The exact wavelength of the fluorescence observed (λ_{em} must be longer than 400 nm to be visible) depends on the substance which emits it: porphyrins fluoresce in the red, pteridine (a fungal metabolite) in the blue green, and some proteins in the blue white. All these compounds absorb (λ_{abs} around 370 nm) and are photochemically excited by "blacklight".

Finally, light traps are useful tools in insect control programs. Their use is based on the observation that insects are attracted to light, the response being dependent on species and wavelength. Kirkpatrick et al. (1970) found that the Indianmeal moth preferred green light rather than ultraviolet and green light together. Ultraviolet is also a common insect attractant. Brower and Cline (1984) observed that significantly more *Trichogramma pretiosum* and *Trichogramma evanescens* responded to ultraviolet than to white light traps. Stuben (1973) reported that ultraviolet radiation of 365 nm flashes lasting 0.0002 s significantly increased the mortality of *Musca domestica*. Bruce (1975) noted differences in the sensitivity of Indianmeal moth eggs of different ages to ultraviolet radiation. Both the larvae and pupae were also sensitive to ultraviolet exposure.

2.6.3 Photodegradation of Foods

Food chemists have been aware of the detrimental effects of light on foods for a long time. The problem has, however, became more acute in contemporary retail with large, self-service, well-illuminated supermarkets. The modern marketing policy involves the display of many foods in transparent and translucent packaging under attractive, high-intensity fluorescent light.

Foods vary in their sensitivity to light but many foods are light sensitive. Colored and moist foods, and those containing high concentrations of unsaturated lipids, are particularly light sensitive. A great deal of beverages and food products are degraded via sensitized reactions (Spikes 1981; Rosenthal 1985).

The light-sensitive components of foods can be divided into two classes: initiators of photochemical reactions and substrates for these reactions (Table 11). In the first category are molecules which contain a suitable chromophore, absorb light and are capable of inducing a photochemical process. The second group comprises materials which are sensitive to the "dark" reactions following the photochemical activation of the sensitizer and are, basically, molecules susceptible to free radical attack. These compounds are, in most cases, the propagators of the photochemical degradation. In a few instances, when the dark reaction yields a sluggish free radical (like phenolic antioxidants), the substrate acts as an inhibitor of the photochemical process. Since the wholesomeness of food is very sensitive to

Light Absorbers and Photoinitiators in Foods 89

Table 11. Photochemical initiators and substrates in foods

Photochemical initiators in food	Substrates of photochemical reactions
Riboflavin (vitamin B_2)	Unsaturated fatty acids
Red no. 3 (erythrosine)	Carotenes
Tryptophane	Vitamins A, D, E, folic acid
Cyanocobalamin (vitamin B_{12})	Other lipids, cholesterol, phospholipids
Vitamin K	Methionine, cysteine
Chlorophylls	Ascorbic acid
Anthocyanines, betalains	
Flavonoids	
Heme pigments	
Tannins	

minute chemical changes in color or flavor, even photochemical processes with a very low overall efficiency are relevant.

2.6.4 Light Absorbers and Photoinitiators in Foods

From an optical point of view foods are complicated physicochemical matrices. Their physical state, spanning from liquids, emulsions, or suspensions to solids, and the complexity of chemical composition means that incident light is not only partially reflected or transmitted, but also scattered and absorbed. The penetration depth in most foods is restricted to few millimeters (Birth 1978). Any molecule which absorbs the incident radiation can, in principle, initiate a photochemical reaction. Therefore, any colored material is susceptible, in time, to photochemical modifications. Naturally, the food colors are the main absorbers in the visible range. Among the natural pigments, porphyrins of plant or animal origin, such as chlorophylls or hemes, are well-known singlet oxygen sensitizers (Foote 1976). Thus, the photooxidation of vegetable oils and fats is sensitized by natural tetrapyrrole derivatives; the photolability of virgin olive oil, which is marketed unrefined as a greenish liquid, was related to the content of chlorophyll (Fedeli and Brillo 1975). Other natural colors, such as betalains and anthocyanines, have been less investigated as photosensitizers but are sensitive to light and oxygen. Thus, anthocyanines from cranberry, red cabbage (Sapers et al. 1981) grape (Palamides and Markakis 1975), and blueberry (Simard et al. 1982) as well as betanine, the red pigment in beets, (Attoe and von Elbe 1981) were reported to selfsensitize their photooxidation.

Finally, a most common natural photosensitizer is riboflavin. Although riboflavin is not a dominant food color, since its tinctorial potency is low (the greenish color of whey is due to riboflavin), its absorptions in the near ultraviolet-visible range of the spectrum (λ_{max} = 371, 444, 475 nm) are large

90 Ultraviolet-Visible Radiation

enough to initiate a variety of photochemical reactions. Photoexcited riboflavin initiates preferentially free radical oxidations, although it might also produce some singlet molecular oxygen (Foote 1976). Riboflavin, photoexcited by exposure to fluorescent light, is photoreduced by readily oxidizable substances, like serum proteins of bovine milk. The reduced riboflavin reacts in turn with molecular oxygen to regenerate the original pigment and produce the superoxide radical anion (Korycka-Dahl and Richardson 1978). The riboflavin-sensitized photooxidation of methyl linoleate and methyl oleate yielded the same isomers of hydroperoxides as obtained by autoxidation; this indicated a free radical mechanism (Chan 1977).

A very significant category of natural colors, carotenes, are unable to sensitize photooxidations. The low-lying triplet excited state of carotenes is below the energy level of singlet oxygen, making the energy transfer to molecular oxygen endothermic, and therefore highly inefficient. Excited carotenes (always of $\pi \rightarrow \pi^*$ type transition) lack also the ability to abstract hydrogen atoms and initiate free radical oxidations. As a matter of fact, carotenes, like all other unsaturated lipids, are compounds prone to free radical autoxidation. Carotenes were also among the first class of compounds to be identified as physical quenchers of singlet oxygen. Since the energy transfer from singlet oxygen to β-carotene is exothermic, it occurs at a diffusion-controlled rate to give ground-state oxygen and unchanged β-carotene. The excess energy is thermally dissipated through cis-trans isomerizations in the polyene chain:

$$^1O_2 + \beta\text{-carotene} \rightarrow {}^3O_2 + \beta\text{-carotene}.$$

The ability of carotenes to act as chemical substrates for photochemical reactions or, alternatively, to deactivate singlet oxygen, i.e., quenching rate, depends on the length of the conjugated chain; the diffusion-controlled limit for carotenes is approached with 11 or more conjugated double bonds (Mathews-Roth et al. 1974). For these reasons, carotenes may exhibit a stabilizing or destabilizing effect with regard to oxidative damage, depending on the specific physical and chemical parameters of the system.

Among the synthetic food colors, erythrosine (red no. 3) is the only efficient singlet oxygen sensitizer. Forty-three certified food, drug, and cosmetic colors were screened for chemical photosensitizing activity by monitoring the conversion of 2,2,6,6-tetramethyl-4-piperidone to the N-oxide radical, which reflects singlet oxygen production. Only xanthene-type dyes substituted with bromine or iodine were found to generate singlet oxygen in this test (Rosenthal et al. 1988).

2.6.5 Constituents of Foods Sensitive to Photodegradation

Oils and Fats. The fatty acid constituents of oils and fats are most sensitive targets for photooxidation of foods. This is due to the presence of a large

Constituents of Foods Sensitive to Photodegradation 91

number of double bonds which are susceptible to the addition of free radicals, the existence of allylic positions prone to hydrogen atom abstractions, and the better solubility of molecular oxygen in the lipid phase as compared to the aqueous phase.

Oxidation of lipids can occur in foods containing substantial amounts of fat, like milk and meat products, oils, and nuts and also in those which contain only minor amounts of lipids, such as vegetable products. For example, soybean oil is particularly susceptible to oxidation due to its high concentration of linoleic acid. The chlorophyll-sensitized photooxidation of soybean oil is inhibited by β-carotene which acts as a singlet oxygen quencher (Lee and Min 1988). Also, exposure of ground pork and turkey to light from "cool light" fluorescent tubes resulted in a higher content of peroxides than in samples kept in the dark. This oxidative effect of light was attenuated by the presence of singlet oxygen quenchers or free radical scavengers, such as butyl hydroxyanisole (BHA). Dissociated hematin and especially the protoporphyrin IX ring, possibly released or exposed by grinding action, were the presumed photosensitizers in this system (Whang and Peng 1988).

Practically all quality attributes of food can be affected by oxidation. Thus, alterations in aroma result from formation of volatile odorous compounds, modifications in taste are caused by formation of new hydroxy acids, color darkening results from condensation reactions between oxidation products and proteins, and, finally, a new texture might be attributed to the oxidative induction of protein crosslinks. Not unexpectedly, the nutritive value and safety of food are impaired.

Classical studies have established the mechanism of autoxidation of lipids as a free radical chain reaction which involves three stages: initiation, propagation, and termination (Pryor 1976) (Fig. 15).

The initiation step is the most intriguing aspect of this chemical process. The abstraction of a hydrogen atom from a fatty acid, particularly an allylic one, Eq. (1), is a facile reaction which explains the rapid degradation of

$$
\begin{align}
\text{INITIATION} \quad & RH + ? & \rightarrow & \quad R\bullet & (1) \\
& R' - CH{=}CH\text{ -}R'' + O_2 \rightarrow \text{transient} & \rightarrow & ROOH & (2) \\[1em]
\text{PROPAGATION} \quad & R\bullet + O_2 & \rightarrow & \quad ROO\bullet & (3) \\
& ROO\bullet + RH & \rightarrow & \quad ROOH + R\bullet & (4) \\
& ROOH & \rightarrow & \quad RO\bullet + \ OH\bullet & (5) \\[1em]
\text{TERMINATION} \quad & R\bullet + R\bullet & \rightarrow & \text{ Nonradical products} & (6) \\
& \text{Other free radical rearrangements}
\end{align}
$$

Fig. 15. The autoxidation reaction

fatty foods exposed to free radicals. In principle, any free radical present in the food matrix can initiate the oxidation of the fatty component and the photochemical reactions which are free radical reactions par excellence can easily do it. Alternatively, the direct addition of an oxygen molecule to a double bond to generate hydroperoxide compounds, Eq. (2), is prevented by quantic restrictions imposed by spin conservation. The ground-state oxygen, which is in the triplet state, cannot add to another ground-state molecule in the singlet state to yield a product also in the ground-singlet state. This spin restriction may be removed by electron excitation of ground-state oxygen to excited-singlet oxygen. Indeed, singlet oxygen may react a thousand times faster than ground-state oxygen with some unsaturated sites of lipids to give hydroperoxides. In foods, the most feasible source of singlet oxygen is photosensitization by pigments, either natural or artificial.

Once a free radical is generated, the chain reaction of oxidation is initiated, and new free radicals, carbon- and oxygen-centered, are formed, and the process is easily propagated. The net chemical result of lipid oxidation is very complex (Frankel 1985). Multiple initial products result from one starting material and many more are generated by the decomposition of unstable hydroperoxides. This decomposition proceeds by homolytic cleavage of peroxy bonds to form alcoxy radicals. These radicals undergo carbon-carbon cleavage to form breakdown products including aldehydes, ketones, alcohols, hydrocarbons, esters, furans, and lactones. Lipid hydroperoxides can react again with oxygen to form such secondary products as epoxyhydroperoxides, ketohydroperoxides, dihydroperoxides, cyclic peroxides, and bicyclic endoperoxides. These secondary products can, in turn, decompose like monohydroperoxides to form volatile breakdown products. Alternatively, the hydroperoxides can condense into dimers and polymers.

The oxidation products of lipids are highly reactive compounds and can also interact with proteins. In food systems, the functions of lipoproteins include stabilization of dispersions against coalescence and participation in texture formation. Consequently, changes in lipoproteins, for instance, due to oxidation of the lipid moiety, will affect the texture of liquid or solid foods. Damage to proteins, initiated by lipid peroxidation, has been observed in model laboratory systems and strongly suggests that similar interaction between peroxidizing lipids and proteins may also occur in the matrices of foods. This interaction may proceed along two different chemical avenues. In one process, transient intermediates in lipid oxidation, such as free radicals and hydroperoxides, react with proteins. This reaction yields protein-centered free radicals and results in subsequent structural alterations. Alternatively, proteins react with stable lipid oxidation products, such as aldehydes, ketones, epoxides, or oximes, to yield new products or crosslinks between protein chains. In particular, bifunctional secondary products of lipid oxidation, such as malondialdehyde, are powerful crosslinking agents which react with amino groups of enzymes and proteins. It is

Constituents of Foods Sensitive to Photodegradation

expected that similar chemical interactions might affect any other food components such as vitamins and carbohydrates.

Fatty acids are not the only lipids susceptible to singlet oxygen attack. Cholesterol is also readily oxidized by singlet oxygen to yield 3β-hydroxy-5α-hydroperoxy-Δ^6-cholestane, which, in turn, decomposes and can initiate free radical chain oxidation of unsaturated fatty acids (Doleiden et al. 1974). Five photooxidation products of cholesterol were identified in spray-dried yolk exposed to fluorescent light or sunlight: 7-ketocholesterol, 7α- and 7β-hydroxycholesterols, cholesterol-$5\beta,6\beta$-oxide, and cholestane-$3\beta,6\alpha$, 6β-triol (Chicoye et al. 1968). In butter exposed to fluorescent light, 5-chlosten-$3\beta,7\alpha$-diol and its 7β-epimer, products of free radical attack, were detected among the oxidation products along with a compound suspected to be 6-cholesten-$3\beta,5\alpha$-diol. This latter compound forms only following singlet oxygen attack on cholesterol (Luby et al. 1986).

Oxidative reactions of foods can result not only in destruction of valuable nutrients and in production of undesirable flavors and odors affecting the palatability of foods but also in generation of toxic compounds. Thus, some oxidation products of cholesterol are angiotoxic (Peng et al. 1976) or carcinogenic (Smith and Kalig 1975).

Since the acceptability of a food product depends on the extent to which deterioration has occurred, some criteria for assessing the extent of oxidation are required. Obviously, sensory analysis is the most sensitive and direct method, but it is not practical for quantitative routine analysis. Consequently, many chemical methods have been developed to quantify oxidative deterioration with development of unpalatable flavors. Table 12 summarizes a few of the assays available for measuring the extent of lipid peroxidation.

In spite of the multitude of assays a universal method which correlates well with the extent of food deterioration throughout the entire course of autoxidation is not yet available. The present methods give information about particular stages of the oxidation process and some are more applicable to certain lipid systems than are the others. This situation should not be unexpected in view of the chemical diversity of food matrices and of oxidation pathways.

When the process of oxidation is followed comparatively with time, the loss of lipid substrate, or the amount of oxygen uptake, it can serve as a general, though nonspecific and not usually very sensitive, index for peroxidation of lipids. This type of test is primarily used to evaluate the susceptibility of a fatty food under conditions which favor oxidative rancidity.

Since the primary products of lipid oxidation are hydroperoxides, it is reasonable to determine their concentration as a measure of oxidation. The "peroxide value" test reflects the total concentration of peroxides and hydroperoxides present at a certain time. However, this approach is restricted by the chemical instability of these compounds; after their con-

94 Ultraviolet-Visible Radiation

Table 12. Compendium of assays for lipid oxidation

Monitored reaction effect	Method of assay
Loss of lipid substrate	Gas chromatography (Slater 1984)
Oxygen uptake during oxidation	Oxygen uptake (Slater 1984)
Formation of peroxides and hydroperoxides	Iodometry (Pryor and Castle 1984); enzymatically (Heath and Tappel 1976); chemiluminescence (Tamamoto et al. 1987)
Formation of malondialdehyde	Direct determination by UV absorption (λ_{max} = 245 nm) or derivatization with thiobarbituric acid (absorption at λ_{max} = 532 nm) or fluorescence at λ_{em} = 553 nm) (Bird and Draper 1984); HPLC (Esterbauer et al. 1984); fluorescence of Schiff base (λ_{em} = 455 nm) (Dillard and Tappel 1984)
Formation of conjugated dienes	Increase in OD at λ = 233 nm (Recknagel and Glende 1984)
Formation of carbonyl compounds	Direct determination by GS-MS, HPLC, derivatization with 2,4-dinitrophenyl hydrazine (Esterbauer and Zollner 1989)
Formation of free fatty acids	Titration, electric conductivity (Laubli et al. 1986)

centration reaches a maximum level it decays as a function of temperature, with the presence of other food components, etc.

An alternative approach to the determination of the extent of oxidation is the measurement of products of hydroperoxide degradation. In contradistinction to the peroxide determination, such an assay is not limited to the early stages of oxidation and may reflect the formation of products, like carbonyl compounds, which actually contribute to the rancid and other objectionable organoleptic flavors. The application of methods based on this approach requires a detailed knowledge of the chemistry involved, stability of the compounds assayed, etc. Among these methods the thiobarbituric acid (TBA) test is one of the most common, in spite of the criticisms of its reproducibility and reliability. The test is based on the color product resulting from the condensation of TBA with malonaldehyde, which is presumably generated in the oxidized fats. However, a large body of evidence suggests that other food components can react with TBA to generate the same chromophore, and even the formation of malonaldehyde is dependent on the composition of the initial lipid.

The assays of oxiranes, conjugated dienes, trienes, aldehydes, and fluorescent compounds are some of the many other assays based on quantitation of suspected end-products of lipid oxidation.

Constituents of Foods Sensitive to Photodegradation

Proteins. Aromatic amino acids (tryptophane, phenylalanine, and tyrosine), histidine, and sulfur-containing amino acids (cysteine, cystine, and methionine) absorb, and consequently can be chemically modified by, short ultraviolet radiation ($\lambda = 254\,nm$). When these amino acids are incorporated in a protein chain, these oxidative changes may interfere with the natural conformation of the protein. This is expressed in denaturation, aggregation, or cleavage, as well as changes in viscosity, surface tension, solubility, or optical or electric properties.

Five amino acids that occur in typical proteins, cysteine, histidine, methionine, tryptophane, and tyrosine, and all of which have electron-rich side chains, are liable to photosensitized oxidation (Schaich 1980b). Therefore, any protein-rich food can be expected to be sensitive to light and oxygen. Most of the experimental observations reported in the literature are, however, on milk and dairy products (Sattar and de Man 1975; Spikes 1981; Rosenthal 1985). In addition to the "oxidized" flavor of milk overexposed to light and which is caused by oxidation of fatty acids, a distinct type of unpleasant flavor, described as "burnt", "scorched", "cabbage", "mushroom", or "activated", is attributed to the photooxidation of proteins. Milk riboflavin can photosensitize the oxidation of methionine to 3-methyl mercapto propionaldehyde (methional) and this imparts to the milk a flavor very much like the solar "activated" unpalatable flavor. A protein with this typical flavor was isolated from whey, indicating that serum proteins are the target of photomodification. Since the milk coagulation time by rennet is also increased in illuminated milk, it was suggested that this was due to riboflavin-mediated oxidation of methionine to methionine sulfoxide. A methionine residue borders the rennet-sensitive bond of κ-casein, which is the protein responsible for the stability of casein micelles. The oxidation of this methionine residue reduces either the accessibility of the protease enzyme or the bond lability, or both. However, methional is not the exclusive photoproduct derived from methionine. While photosensitization by flavines yields methional, methionine is oxidized by most sensitizing dyes to the sulfoxide. This diversion in product is probably caused by a change from type II to type I with flavines (Foote 1976).

Probably, methionine is not the only amino acid in milk that is affected by light and riboflavin. The exposure of milk serum to fluorescent light chemically modified an other ten amino acid residues in immunoglobulins. The photooxidation of tryptophane residues by light of $\lambda_{max} = 450\,nm$ was also suspected of yielding products responsible for the unpleasant flavor of milk. Cystine residues in milk are degraded by light to mercaptans, sulfides, and disulfides. Finally, histidine, tryptophane, and tyrosine residues were destroyed by the photooxidation of β-casein.

In simple, model systems, N-formylkynourenine could be isolated as a primary product of photooxidation of tryptophane but it is, however, readily converted to complex mixtures of products. Tyrosine, which has also been studied extensively, shows reactivity to a type I mechanism. The products are not known, but the reaction leads to the cleavage of the phenolic ring. Cysteine is slowly oxidized to cystine and, under certain conditions, to cysteic acid. Histidine is one of the most reactive amino acids with singlet oxygen and gives products of cleavage of the imidazole ring; the initial products have not been isolated, but model studies suggest that cleavage of the enamine double bond is the likely primary result. The kinetics and mechanisms of these reactions depend on the amino acid, the sensitizer, the nature of the medium, the pH, and the oxygen concentration (Spikes 1989).

An important technological property of muscle proteins for manufacture of gel-type foods is the thermal gelation ability. The denaturation of muscle proteins is the underlaying process of gelation of meat paste. Ultraviolet irradiation of sardine, beef, or pork paste, prior to thermal gelation, affects the gel strength. When ultraviolet irradiation $(2700 \mu \, W/cm^2)$ from a mercury arc lamp was applied to meat pastes, the surface strength of thermal gels markedly increased with irradiation time. The effect of 360 nm radiation on the sardine gels was superior to that of 250 nm wavelength. Actomyosin ATPase is denaturated by ultraviolet radiation, with simultaneous activation of Mg-ATPase and decrease of EDTA-ATPase. The activation of Mg-ATPase seems to be responsible for an even stronger interaction between myosin and actin with a consequent enhancement of the gel strength (Tuguchi et al. 1989).

Conformational effects may make some residues in macromolecules less susceptible to photooxidation than others. Photooxidizable residues exposed at the surface of the protein molecule are degraded more rapidly than residues located in the interior of the molecule; these differences often disappear on denaturation. Directly related to the photodegradation of proteins is photodeactivation of enzymes in foods. Destruction of key active-site amino acids, most often methionine and histidine, or damage leading to loss of conformational stability causes inactivation of many enzymes. Thus, lipase in milk lost 80% of the initial activity after 30 min exposure to sunlight, apparently due to photosensitization by riboflavin.

Carbohydrates. The simple carbohydrates do not absorb in the ultraviolet-visible region above 300 nm and are also inert to singlet oxygen attack. Any possible photodegradation of sugars, like depolymerization of polysaccharides, is due to secondary, free radical reactions. Indeed, sugars and polysaccharides are photooxidized fairly readily in the presence of ketonic sensitizers. These are sensitizers of n $\rightarrow$ π^* type, which most likely initiate type I photooxidations by hydrogen atom abstraction. Indeed, they abstract a hydrogen from the α-carbon of the alcohol group giving an alcohol radical; this, in turn, reacts with oxygen to yield carboxylic acids, ketones, and aldehydes. This formation of free radicals prompts also rearrangements which result in chain scission giving depolymerized products.

Constituents of Foods Sensitive to Photodegradation

Vitamins. Probably the most significant aspect of the photooxidative degradation of foods from a nutritional viewpoint is the destruction of vitamins. These vital compounds, which belong to diverse chemical classes and are found in minute amounts in food products, are very likely targets of photooxygenation. All vitamins, in spite of the chemical diversity, contain double bonds which are susceptible to oxidation and free radical attack. Vitamins A, B_2 (riboflavin), B_6 (pyridoxine), folic acid, B_{12}, C (ascorbic acid), D, and E (tocopherol) have all been reported to be destroyed in various foods by light and oxygen. The extent of destruction depends on the nature of the food and the conditions of illumination. Again, most studies have dealt with the photooxidation of milk when sunlight or fluorescent light was the illumination source (Sattar and de Man 1975; Spikes 1981; Rosenthal 1985).

From the specific viewpoint of food photochemistry, two vitamins are particularly interesting: riboflavin (Fig. 16) and tocopherol (Fig. 17). Riboflavin is a most common natural light-absorbing species in foods. Tocopherol, on the other hand, has a significant singlet oxygen quenching ability (Foote et al. 1978) and might serve as an antiphotooxidant.

Riboflavin was found to be very photolabile in solution and the rate of light-induced degradation followed first-order kinetics and increased with increasing temperature. In milk, the decomposition reaction rate was slightly

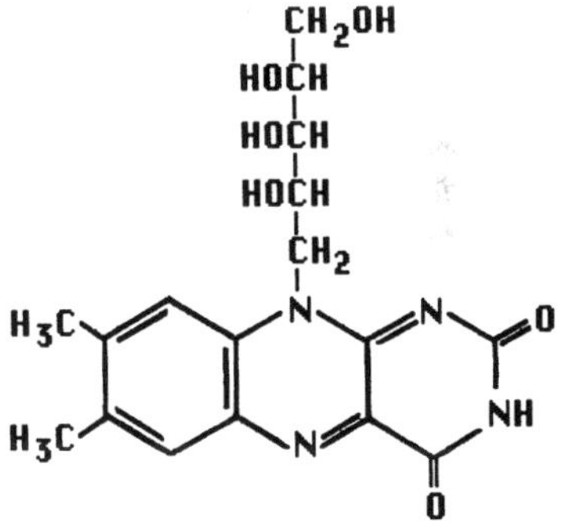

Fig. 16. The chemical structure of riboflavin

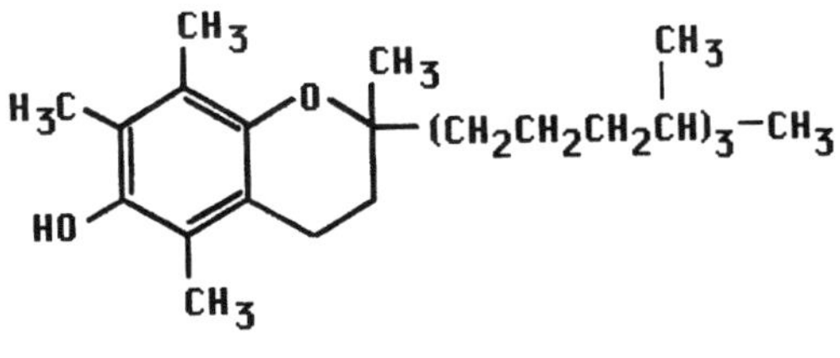

Fig. 17. The chemical structure of tocopherol

98 Ultraviolet-Visible Radiation

dependent on the milk composition: $1.86 \times 10^{-5}\,s^{-1}$ in skimmed milk and $1.47 \times 10^{-5}\,s^{-1}$ in whole milk (Allen and Parks 1979). Riboflavin within the food matrix is also most light sensitive. Retention of riboflavin in light-exposed, continously agitated milk was significantly lower than in quiescent samples. Sampling positions within translucent polyethylene milk containers also had a significant effect on light-induced riboflavin loss. After 5 days of light exposure, riboflavin retention at the top of the still milk containers averaged 58% compared with 92% retention for milk located at the bottom of the containers. Opaque, aluminum-vapor-coated, plastic film used as overwraps decreased the penetration of visible light into milk samples and consequently reduced the riboflavin degradation (Palanuk et al. 1988).

As already mentioned, visible light affects the flavor of milk due to a riboflavin-sensitized modification of proteins. Riboflavin can also photosensitize the destruction of other vitamins, such as vitamin C and folic acid.

Tocopherols are very sensitive to oxidation by singlet oxygen, yielding a variety of products that are solvent dependent. The rate constants of singlet oxygen quenching by tocopherols in ethanol were determined as follows: k_q (α-tocopherol) $= 2.6 \times 10^8\,M^{-1}\,s^{-1}$, k_q (γ-tocopherol) $= 1.8 \times 10^8\,M^{-1}\,s^{-1}$, and k_q (δ-tocopherol) $= 1.0 \times 10^8\,M^{-1}\,s^{-1}$. On the other hand, the chemical reaction rates were: k_r (α-tocopherol) $= 6.6 \times 10^6\,M^{-1}\,s^{-1}$, k_r (γ-tocopherol) $= 2.6 \times 10^6\,M^{-1}\,s^{-1}$, and k_r (δ-tocopherol) $= 0.7 \times 10^6\,M^{-1}\,s^{-1}$. Because of their high chemical reactivity toward singlet oxygen, tocopherols are probably poor photoprotectors in food, although in a model system they can show an inhibitory effect.

Other Light-Sensitive Food Constituents. In addition to lipids, proteins, and vitamins other food constituents may be affected by light. Although minor components, their modification can be crucial for the quality of a certain product.

Beer. For over a century it has been known that beer develops an unpleasant flavor on exposure to daylight, described as "light-struck" flavor. For this reason, the brown bottle was the container of choice for beer (until the appearance of aluminum cans). There is convincing evidence that the "sun" taste is associated with the presence of iso-α-acids in beer. The iso-α-acids are formed by a rearrangement of the α-acids during the boiling of wort in the beer brewing process (Fig. 18).

Iso-humulone [R $= -CH_2CH(CH_3)_2$] is the most prominent representative of iso-α-acids and forms the basis of the "bittering principle" and of the typical beer flavor.

The riboflavin or other pigments in malt are the actual light-absorbing species which most probably sensitize a photochemical reaction of iso-α-acids to yield unpleasant flavor substances. The unsaturated iso-hexenoyl

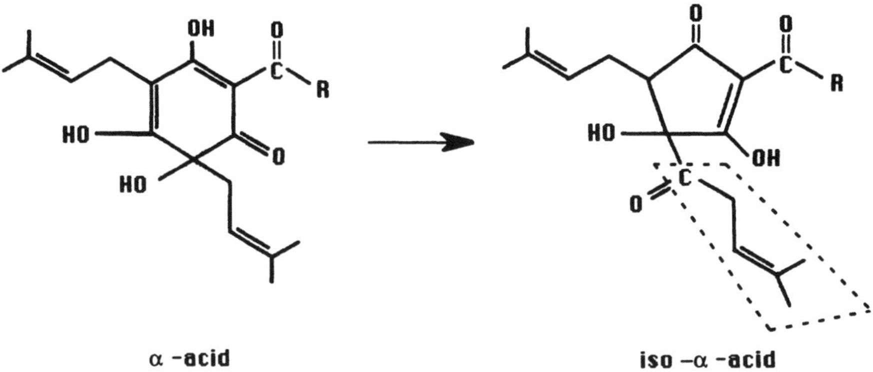

Constituents of Foods Sensitive to Photodegradation

Fig. 18. The chemical structure of α-acid

side chain of the iso-α-acids seems to play an importnat role, since no "sun" flavors develop when this unsaturated side chain is hydrogenated catalytically or by sodium borohydride, $NaBH_4$. As a matter of fact, the hydrogenation treatment is carried out on an industrial scale for production of a beer which is sold in colorless, transparent bottles.

The photodegradation of beer is one of the few light-induced reactions which are inhibited by oxygen. However, saturation of beer with oxygen cannot be used to avoid the development of "light-struck" flavor since the quality of such a beer deteriorates during storage in the dark (Vogler and Kunkley 1982).

Potatoes. The postharvest exposure of potatoe tubers to natural or artificial yellow and red light causes "greening" of the surface, an undesired process caused by the induction of chlorophyll production. Certain wavelengths, in the blue region, induce also the less visible solanine formation. The presence of this alkaloid is responsible for the impairment of flavor and a bitter taste (Smith 1968).

A compendium of some reports on food products deteriorated by light is presented in Table 13.

The most effective method of preventing photochemical degradation is to package the food products in opaque materials. Vacuum packaging in impermeable, transparent films is also extensively used. Next to it is the use of transparent but colored materials which reduce the total light intensity inside the package or selectively remove harmful wavelengths. The same effect can alternatively be achieved by using properly filtered light for illumination of foods on display. However, the practice of employing colored film food packages or colored light, which unavoidably results in change of the product appearance, may be occasionally considered deceptive or unattractive.

100   Ultraviolet-Visible Radiation

Table 13. Food products deteriorated by light

Food product	Cause of deterioration	References
Vegetable oil	Oxidation	Fedeli and Brillo (1975), Sattar et al. (1976)
Cream, butter, butter oil	"Off-flavor", oxidation of cholesterol, decomposition of vitamin A and β-carotene	Foley et al. (1971, 1977), Sattar et al. (1977), Luby et al. (1986)
Yogurt	Oxidation	Bosset et al. (1986)
Milk	Light-induced flavor, generation of volatiles, loss of riboflavin and vitamin A	Sattar and deMan (1975), Bray et al. (1977), Allen and Parks (1979), Gilmore and Dimick (1979), Hoskin and Dimik (1979), Methta and Bassette (1979), deMan (1980), Gaylord et al. (1986), Toyosaki et al. (1988), Bartholomew and Ogden (1990)
Meat, meat products	Changes in color, rancidity	Watts (1954), Bailey et al. (1964), Solberg and Franke (1971), Setser et al. (1973), Zachariah and Satterlee (1973), Satterlee and Hansmeyer (1974), Bala and Nauman (1977), Chan et al. (1977)
Rice	Rancidity	Sowbhagya and Bhattacharya (1976)
Walnuts, pecan nuts	Color change, rancidity	Musco and Cruess (1954), Heaton and Shewfelt (1976)
Frozen peas	Flavor and color changes	Sheperd (1959)
Dried fruits	Color changes	Bolin et al. (1964), Bothill and Hawker (1970)
Sliced potatoes	Increase in solanine content	Salunkhe et al. (1972)
Potato chips	Rancidity	Quast and Kare (1972), Chan et al. (1978)
Infant liquid formula	Degradation of vitamin C	Mack et al. (1976), Singh et al. (1976)
Frozen orange drinks	"Off-flavor", loss of vitamin C	Ahmed et al. (1976)
Beer	"Off-flavor"	Vogler and Kunkley (1982)
Baked rolls	Loss of riboflavin	Loy et al. (1951), Stephens and Chastain (1959)
Sauerkraut	Pink-greyish discoloration and "off-flavors"	Steinbuch and Rol (1986)
Cinnamon oil	Oxidation	Palamides and Markakis (1975), Carnevale et al. (1980), Attoe and von Elbe (1981), Kearsley and Rodriguez (1981), Sapers et al. (1981), Simard et al. (1982), Hicks and Abdullah (1987)
Vegetable pigments (betanine, anthocyanines, carotenoids)	Color degradation	

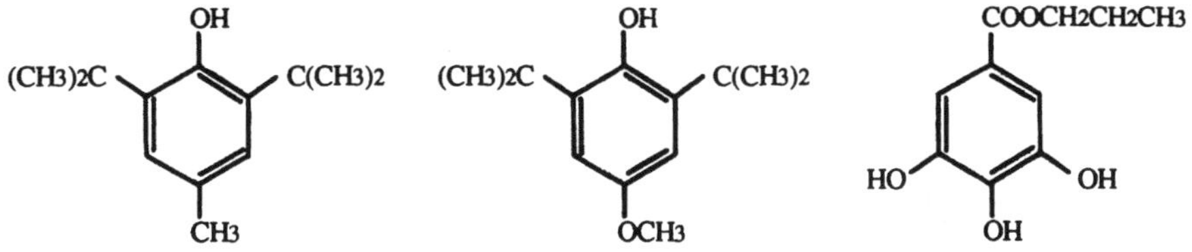

Applications of Ultraviolet-Visible Radiation in Food Analysis

3,5 - Bis (t - butyl) -
4 - hydroxytoluene
BHT

3,5 - Bis (t - butyl) -
4 - hydroxyanisole
BHA

Propyl gallate

Fig. 19. Chemical structures of synthetic antioxidants

The common synthetic antioxidants, butyl hydroxyanisole (BHA), butyl hydroxytoluene (BHT), and propyl gallate (Fig. 19), although primarily designed for the inhibition of thermal oxidation, could be also useful for minimization of photodegradation. These compounds are good scavengers of reactive free radicals, which would otherwise freely propagate the chain reactions responsible for the major damage initiated by light.

2.7 Applications of Ultraviolet-Visible Radiation in Food Analysis

The interaction of ultraviolet-visible radiation with matter is the basis of the most common analytical technique, that is, ultraviolet-visible spectroscopy. This is the tool for accurate color measurements, for assays of major food constituents, micronutrients, food additives, and food contaminants and for analysis of water, etc.

Color is a very common index for grading foods and is often used as a determinant of the ripeness of fruits, as one of the most prominent variables apparent in raw or cured meats, and as an index of efficiency of wheat cleaning prior to milling, etc. Precise measurement of color in foods is performed instrumentally. Spectroscopic equipment has been designed to measure directly the color of agricultural products, either by spectral absorbance of transparent samples or by spectral reflectance from opaque samples. Also, chemical extraction of pigments from foods and measurements of their concentration at specific wavelengths in a spectrophotometer, can, within certain limitations, be satisfactorily correlated with the color impression. Not to be forgotten, the color of the outer surface, and not necessarily the total pigment content, is responsible, in general, for color impression.

The absolute concentration of a food constituent is accurately determined from its absorption or emission spectrum. The absorption spectrum in the ultraviolet-visible region is usually a plot of wavelength of absorption vs the absorption intensity (expressed either as absorbance or transmittance). The principal characteristics of an absorption band are its position and intensity. The position of an absorption band depends on the chemical structure of the compound assayed and corresponds to the wavelength of energy required for its characteristic electronic transitions; the intensity of the absorption band is a function of the concentration of the absorbing compound and is conveniently established by the Beer-Lambert law. Since the extinction coefficient for a certain molecule is known, or can be easily determined, measuring the absorbance provides an easy way of revealing the concentration.

Fluorescence and phosphorescence spectra are also specific features of a chemical structure and can be used for identification and quantitation. Fluorescence measurements are among the most sensitive methods of analytical chemistry. The fungal poisons, aflatoxins, after chromatographic separation, can be assayed by fluorescence in amounts lower than nanograms. Nonfluorescent compounds are often chemically derivatized to fluorescers, to increase the sensitivity and facilitate detection after chromatographic separation. On the other hand, the application of phosphorescence in food analysis is still very limited.

Related to fluorescence and phosphorescence is the phenomenon of delayed light emission. All green plants, when irradiated with visible light, give off light for a period of several seconds after illumination. This phenomenon of delayed light emission was explained as a reversion of certain reactions of photosynthesis which are capable of releasing a portion of their stored chemical energy through chemiluminescence. Measurements of delayed light emission have focused so far on maturity evaluation of fruits because there is a definite relationship between the emitted light and chlorophyll concentration, which changes with maturity (Gunasekaran 1990).

In recent years, the possibility of integral nondestructive testing has received a lot of attention. Development of a nondestructive quality test allows instantaneous measurement of desired chemical characteristics and depends entirely upon establishing a relationship between the quality factor of interest and a physical property of the sample. It commonly involves application of a well-characterized source of energy to the test sample. Depending on the effect monitored, the energy may come from any part of the electromagnetic spectrum range: X-rays, ultraviolet, visible, infrared, or microwaves (it should be noted that ultrasonics are also employed in nondestructive testing). The energy which reaches the sample interacts with it and is modified in some unique manner because of its chemical and physical properties. The incident energy may be scattered, reflected, transmitted, or absorbed. The difference between the input energy and the energy response

Applications of Ultraviolet-Visible Radiation in Food Analysis

Table 14. Nondestructive light-transmission measurements of agricultural products. (Adapted from Finney 1978)

Quality factor	Measurement	Component measured
Maturity, color of apples	Δ_{OD} (690–740 nm)	Chlorophyll
Water core in apples	Δ_{OD} (760–810 nm)	Physical changes in apple tissue
Maturity, color of peaches	Δ_{OD} (700–740 nm)	Chlorophyll
Color of tomatoes	Δ_{OD} (620–670 nm)	Lycopene and chlorophyll
Greenness of oranges	Δ_{OD} (690–740 nm)	Chlorophyll
Maturity of peanuts	Δ_{OD} (490–520 nm)	Carotenoid pigments
Color of cherries	Δ_{OD} (590–620 nm)	Anthocyanine
Blood in eggs	Δ_{OD} (577–597 nm)	Blood
Milling quality of rice	Δ_{OD} (660–850 nm)	Brown bran material
Hollow heart in potatoes	Δ_{OD} (710–800 nm)	Brown material in the vicinity of the void

is measured and recorded and provides the basis for an empirical correlation with product quality. A few nondestructive light-transmission tests of agricultural products are summarized in Table 14.

The use of absorption or emission visible-light spectroscopy is not limited to organic materials. Flame photometry is a spectroscopic technique based on measuring the light emission from excited states of certain inorganic atoms. With the alkaline and alkaline-earth metals, like sodium, potassium, and calcium, the energy required to excite the atoms is relatively low and there is enough energy in a gas-air flame (900–1200 °C) to achieve this. The flame photometer instrument provides a flame whose temperature is hot enough to excite several elements, determines which wavelengths are given off and what their intensities are. The intensity of emission is calibrated for ion concentration.

The development of atomic absorption spectroscopy consitutes an improvement of flame photometry. The instrument is provided with a source of radiation whose wavelength is exactly the same as the atoms in the flame and it measures the decrease in intensity of the radiation from this source, after absorption by the atoms in the flame.

In addition to light absorption or emission, other physical interactions between electromagnetic radiation and matter are employed in food analysis. The amount of light refraction is a characteristic of every substance and is called the refractive index. Refractive index measurements have long been used for the qualitative identification of unknown compounds by comparing the measured values with those tabulated in the literature. Even the refractive index of mixtures could be a revealing parameter because it varies linearly with the mole fraction of the components and can be used to determine the total concentration of the components in a mixture. The determination of total solids in juices and of the concentration of sugar solutions are a few of the common applications of this assay.

Polarimetry, which is based on the measurement of optical rotation, that is, the angle of rotation of the plane of an incident polarized light, enables the determination of optically active compounds. Polarimetric determinations are extensively employed for quantitation of sugars.

An interesting analytical "hybrid" is photoacoustic spectroscopy. In photoacoustic spectroscopy, a sample is placed inside a specially designed closed cell containing air and a sensible microphone. The sample is illuminated with chopped monochromatic light. Any light absorbed by the solid is converted in part, or in whole, into heat by a nonradiative de-excitation process within the solid. The resulting periodic heat flow from the solid absorber to the surrounding gas creates pressure fluctuations in the cell that are detected by the microphone. The technique has been applied in the analysis of some milk products (Martel et al. 1987).

Chapter 3 Infrared Radiation

3.1 The Definition of Infrared Radiation

Infrared radiation refers broadly to that part of the electromagnetic spectrum between the visible and microwave regions, extending approximately over the wavelength range $0.5-100\,\mu m$, where $1\,\mu m = 10^{-6}\,m$. The first scientific report on the heating effect of infrared radiation is believed to belong to Herschel (1800). Knowing that a lens will focus the sun and burn paper, Herschel (the same German-born, English-by-choice astronomer who discovered the planet Uranus in 1781) placed a thermometer in the various colors of the solar spectrum resolved by passing daylight through a prism. He noticed an unusual large heating effect in the region immediately beyond the red end of the visible spectrum (Herschel 1800). This invisible part of the spectrum was naturally called infrared and it was defined by subsequent measurements between wavelengths of 0.76 and $350\,\mu m$.

Infrared radiation is commonly defined by wavelength measured in μm. An alternative parameter, the wavenumber, has also been used as a substitute to frequency.

The wavelength, λ, is reciprocal to the frequency, v:

$$\lambda = \frac{c}{v} \quad \text{or} \quad \frac{1}{\lambda} = \frac{v}{c} = \frac{v}{3 \times 10^{10}}$$

where c is the speed of electromagnetic radiation in $cm\,s^{-1}$.

The term wavenumber, $\bar{v}$, which is still proportional to the energy, is defined as:

$$\bar{v} = \frac{1}{\lambda}$$

and is customarily given in units of cm^{-1} (reciprocal centimeters, sometimes called kaysers). It follows that:

$$\Delta E = h\bar{v}c$$

and the conversion relation between wavelength and wavenumber is:

$$\lambda\,(\mu m) = \frac{10\,000}{\bar{v}\,(cm^{-1})}$$

3.2 Sources of Infrared Radiation

Most light and heat sources, including the sun, give off some energy in the infrared range. An ideal "black body" is, by definition, a surface which absorbs all the energy incident upon it. Consequently, under proper conditions, it should be able to radiate all the energy, namely all wavelengths in a continuous spectrum. As the temperature of a "black body" is elevated, there is a corresponding increase in the amount of energy liberated at every wavelength but not in equal proportions. Thus, the maximum power (W) radiated at a certain wavelength, per unit wavelength interval (μm), divided by unit area (cm^2) of a black body at temperature T is given by:

$$I_{max} = 1.290 \times 10^{-15} T^5 \, W \, cm^{-2} \, \mu m^{-1}.$$

The Stefan-Boltzmann function defines the relationship between the intensity of the integral emitted radiation and the absolute temperature of the source:

$$I = 5.679 \times 10^{-12} T^4 \, W \, cm^{-2}.$$

If a "black body" is hot enough it can emit in the infrared range. Incandescent sources operating at temperatures below about 870 K emit only invisible energy in the infrared. A coal stove or an electric iron are examples of such incandescent radiation bodies. As the temperature of the radiator increases, the energy radiated shifts to shorter wavelengths. It is the shifting of the region of maximum intensity that produces the change in color from red to blue-white as an iron bar is heated. For incandescent sources operating at higher temperatures, like incandescent electrical bulbs, electronic transitions in atoms and molecules result in the emission of visible radiation along with infrared radiation.

The thermal "black body" emission has a bell-shaped distribution of wavelengths spread over the infrared spectrum range and is maximized somewhere between approximately $5 \, \mu$m at 900 K and $2.8 \, \mu$m at 1800 K. A family of curves of "black body" radiators at different temperatures (Fig. 20) illustrates the shift of the crest of the wave toward the region of shorter wavelength with the increase in temperature.

Infrared heating equipment, electric lamps, or gas-fired broilers are specifically engineered so that most of the energy emitted is in the infrared. Typical electric sources are incandescent lamps provided with filaments or ceramic rods. It should be noted that even the common incandescent lamps have a very high output in the infrared and therefore should not be used close to foods that could be damaged by higher temperatures, like in food storage and display areas. This is the reason that fluorescent lamps which have a much lower relative infrared output are extensively used for illumination purposes.

The gas-fired broiler consists of a few, usually three, layers of metal screens sandwiched together. A mixture of air and gas supplied through the

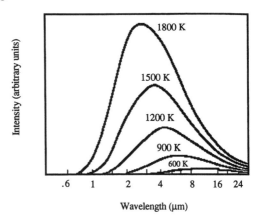

Fig. 20. The emission of a hot "black body"

screens is ignited to produce a blanket of flame about 1 cm thick and of a temperature which varies from 620 to about 800 °C. A temperature of the radiant body of 650 °C produces a maximum energy wavelength of 3.15 μm which is effective in infrared cookery. However, some of the new infrared cooking emitters work at temperatures as high as 950 °C, which is apparently desirable to produce infrared radiation of maximum cooking effectiveness, although the surface of most foods, unless they are very moist, cannot stand temperatures much above 260 °C. In this equipment the food must be placed some distance from the heat source otherwise the food will burn. The amount of energy which reaches food from an infrared source varies geometrically with the distance of the food from the infrared source. Since the propagation of infrared radiation is linear and the depth of penetration is shallow, mechanical provisions are made to rotate large pieces of food in infrared heaters.

3.3 Infrared Heating in Food Processing

The temperature of an object is a measure of thermal motion at the molecular level. As molecular motions become more energetic, the temperature is increased. Eventually, they become large enough to cause changes in physical or chemical structure. To raise the temperature of a food, energy must be supplied to increase the thermal motion of the molecules. In conventional heating, which is achieved by combustion of fuels or by an electric resistive heater, the heat is generated outside of the object to be heated and is conveyed to it by convection of hot air or by thermal conduction. The method of coupling energy from the source to the food distinguishes the type of cooking: baking is by convective heat transfer and frying or boiling by conductive transfer. Energy is applied at, or very near, the surface of the object to be cooked and results in a heated surface region. The object is then heated gradually and progressively from the hot surface toward the inside only by conduction and requires the existence of a permanent temperature gradient. The magnitude of the temperature rise at the center and the elapsed time to reach this temperature depend on the geometry and thermal properties of the food.

In addition to convection and conduction there is a third heating possibility and this is the exposure of the object to radiation which, if absorbed, may be eventually converted into heat; broiling is the conventional type of cooking with radiative transfer. The radiation heating is characterized by freedom from electric and mechanical contact. Any electromagnetic radiation may be degraded into thermal movements of the molecules in the absorbing matter, but the efficiency of this transformation is dramatically dependent on the frequency (or energy) of radiation. Since the energy carried by quanta of radiation at wavelengths shorter than infrared is high enough to induce electronic or chemical changes in the absorbing molecules, like breaking chemical bonds or electronic excitation, the dissipation of absorbed energy as heat is minor. At wavelengths in the infrared range and longer, conversion efficiency of absorbed energy into heat is a major available pathway and, therefore, in practice, radiation heating is generated by the absorption of the electromagnetic radiation in the submillimeter range.

Attempts to utilize the infrared energy of the sun are at least several thousand years old. The burning mirror of Archimedes may well be the most notorious early device. More modest uses in laboratory experiments are documented as early as 1599 when Gesner described "The maner of Distilling in the Sunne" in his monograph (Gesner 1599). Some 50 years later French authored an early record of a culinary implementation of infrared heating: two setups "to rectify spirits" (Fig. 21), which included provisions for the collection and accumulation of solar radiation, were based on the heat capacity of materials such as glass, marble, or cast iron (French 1653).

As with any electromagnetic radiation, infrared is absorbed by organic materials at discrete frequencies corresponding to intramolecular transitions between energy levels. The transitions in the range of infrared energy are expressed by rotational and vibrational (stretching) movements of the interatomic bonds. Therefore, although the infrared absorption spectrum is characteristic of the entire molecule, certain groups of atoms give rise to absorption bands at, or near, the same frequencies, regardless of the structure of the rest of the molecule. Rotational frequencies are in the order of 10^{11}–10^{13} Hz (wavelengths of $30\,\mu$m to 1 mm) and vibrational frequencies are in the range 10^{13}–10^{15} Hz (wavelengths of 0.3 to $30\,\mu$m). In the condensed phase, like liquids, the separations between allowed energy transi-

Infrared Heating in Food Processing

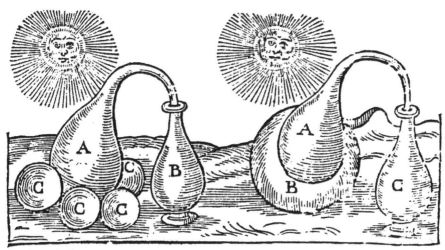

Fig. 21. Two versions of an apparatus for solar distillation of spirits

Table 15. Infrared absorption bands relevant to food heating

Chemical group	Absorption wavelength (μm)	Relevant food component
Hydroxyl group (O — H)	2.7–3.3	Water, carbohydrates
Aliphatic carbon-hydrogen bond	3.25–3.7	Fats, carbohydrates, proteins
Carbonyl group (C = O) (ester)	5.71–5.76	Fats
Carbonyl group (C = O) (amide)	ca. 5.92	Proteins
Nitrogen-hydrogen group (— NH —)	2.83–3.33	Proteins
Carbon-carbon double bond (C = C)	4.44–4.76	Unsaturated fats

tions are very small and, therefore, the absorption of infrared radiation is almost continuous.

The infrared absorption bands characteristic of chemical groups relevant to the heating of food are summarized in Table 15. These strong absorptions are exclusively due to stretching vibrations.

An absorbing substance never becomes saturated with infrared radiation. The vibrationally excited molecules continually lose the energy in random directions by relaxation processes such as molecular collisions, which convey the energy to the surrounding medium as heat.

110   Infrared Radiation

Wavelengths in the range $1.4-5\,\mu m$ are the most effective in cooking food, because they penetrate the blanket of vapors surrounding the food and enter the food to a depth of a few millimeters. Most of the infrared radiation is readily absorbed by thin layers of organic materials or water, and therefore the heating is superficial. On the other hand, infrared heating is fast because the energy is transmitted from the heat source to the food almost instantaneously, not having to be carried to the food by means of hot air as occurs in the case of convection baking. Since the heat is produced on the surface of the material, the interior is heated by conduction and a temperature gradient from the surface to the center always exists. The surrounding air is heated indirectly by the surface of the material and therefore is not quite as hot as in convection or conduction heating.

Comparative assessments were made of the effects of infrared and conventional heating. The cooking losses for infrared broiled steaks were slightly greater than for conventional gas broiled steaks (28 vs 24%), but they had significantly higher moisture and lower fat contents. Thiamin retention was similar for both kind of steaks (*ca.* 81–85%), and no differences were found in maximum shear force values, sensory evaluation scores, or overall acceptability (Vandermey and Khan 1987). Also, the quality of ground-beef patties after infrared heat processing in a conveyorized tube broiler for foodservice use was assessed (Khan and Vandermey 1985). Three protein and two vegetable portion-controlled menu items were monitored for food product yield and nutrient content (four vitamins, three minerals, 18 amino acids, seven fatty acids, and ammonia). Riboflavin and vitamin A levels in hamburgers and tomatoes and total amino acid content in hamburger patties and cod fillets were significantly greater using infrared cooking. No significant differences in minerals were found between the two cooking methods for hamburger patties, baked potatoes, and tomatoes. Nutrient retention for most nutrients analyzed was similar between the infrared and forced-air convection cooking methods (Unklesbay et al. 1983). Sensory quality and energy consumption for baking molasses cookies were compared for infrared, forced-air convection, and conventional deck ovens. The type of oven used for baking made a significant difference in all of the appearance characteristics of the cookies, except for top grain score and number of cracks on the tops. Cookies baked in the convection over had significantly higher mean scores for uniformity in their round shape and for number, uniformity, and depth of surface cracks than did cookies baked in either of the other ovens. Depth of cracks was greatest in the convection oven, less in the infrared, and significantly less in the deck oven. Intensity of the surface browning was least in the convection oven, more in the infrared, and significantly more in the deck oven. Cookies baked in the infrared oven had significantly greater creasing of the upper surface and a greater intensity of interior dark brown molasses color than did cookies baked in either of the other ovens. Mean scores for width of cracks on the

Infrared Heating in Food Processing

top of cookies and for uniformity of brown color were significantly lower for cookies baked in the infrared oven than for those baked in the convection and deck ovens. Energy use in baking was greater in the convection oven than in the other two ovens, but product quality may be worth the added cost (Heist and Cremer 1990). Infrared processing of maize germ and soybeans has also been reported (Kouzeh et al. 1984). Infrared heating in the processing of legume seeds and cereal grains has been reported to gelatinize the starch and increase its availability, to deactivate lipases in oil seeds, and to eliminate antinutritional factors. Although dry beans constitute an important source of inexpensive food protein, their consumption is limited due to association with flatulence and inconvenience created by the "hard-to-cook" phenomenon, which includes long soaking and cooking times. Efforts to improve cooking quality of dry beans have included various irradiation pretreatments with microwaves, gamma and also infrared radiation (Kadir et al. 1990). Although infrared heating of pinto beans to 99 and 107 °C improved the rehydration rate and degree of swelling, the process also increased the cooking time by 25% (after 99 °C treatment) or by 50% (after 107 °C treatment) because of alterations in the structure of the seed or seed coat.

Infrared lamps are commonly used to keep food hot (Cremer and Richman 1987) and to give food a rich red color on buffet tables and cafeteria lines. Another use of infrared lamps is in poultry rearing as a heat source for brooding chicks and turkey poults.

Infrared radiation has been suggested as a method for controlling stored-product insects. Focused solar radiation was found to be suitable for disinfestation of such products as dried fish and grains (Nakayama et al. 1983). Kirkpatrick (1975) demonstrated a 99% death rate of *Sitophilus oryzae* and a 93% death rate of *Rhyzopertha dominica* by infrared radiation which raised the wheat temperature to 48.6 °C. Natural infestations of stored wheat by the weevil *S. oryzae*, the lesser grain borer *R. dominica, Crypolestes pusillus* Schonh, and *Tribolinum castaneum* Hbst. were controlled by raising the temperature to 55 °C (Kirkpatrick et al. 1973). Combining infrared radiation with a partial vacuum (25 torr) is more effective in controlling these pests (Tilton et al. 1983). The practical value of such a treatment remains to be fully investigated.

A traditional and widespread use of infrared radiation is sun-drying of agricultural products. Many food products, ranging from fish to raisins, are dehydrated using solar radiation. This is a process of substantial economic significance in many areas where the weather is reliable in all but freak seasons. The customary technique is to spread the materials to be dried in a thin layer on the ground to expose them to solar radiation and wind. In recent years innovations have been adopted, particularly for fruit drying, in which fruit is placed in carefully designed racks to provide controlled exposure to sun and wind, and to assist with handling. Improved process

112   Infrared Radiation

control and product quality have resulted. Other experiments have centered on crop drying, using either solar-heated air in more or less conventional air dryers, or a combination of direct and air drying by placing the materials to be dried in flat-plate collector-dryers. The sun-drying is occasionally sufficient in itself or is a useful adjunct to dehydration by artificial energy.

A nonconformist example of infrared heating is the preparation of "sun" or "solar" tea, brewed during the summer months in the western part of the United States. This beverage is prepared by perfusing tea-bags with water in a big transparent glass jar exposed to direct sunlight for several hours in the middle of the day. The chemistry of this procedure has not been clarified, but the belief is that precipitates apparently do not form on the subsequent refrigeration of the drink, which qualifies this method as the preferred one for the preparation of iced tea. The infrared – solar heating is also occasionally recommended for soaking fruits in sugar, in folkloristic recipes for jams and marmalades.

3.4 Analytical Applications of Infrared Radiation

The greatest value of infrared spectroscopy undoubtedly lies in its contribution to the elucidation of molecular structure, particularly the recognition of chemical functional groups and their environment. The spectral section of most practical value lies between 1 and $25\,\mu$m. Absorptions in this region are due to the occurrence of vibrational transitions in a molecule within the ground electronic state. This process can be visualized as bonds vibrating with a greater amplitude following the absorption of infrared energy by the molecule. A particular vibrational transition can absorb infrared energy only if that vibration causes a change in the dipole moment of the molecule. Vibration of two identical atoms against each other, for example, in the diatomic oxygen or nitrogen molecules, will not result in a change of the electric symmetry or dipole moment and, therefore, such molecules do not absorb in the infrared region.

Most functional groups have their characteristic absorptions in the limited region between 2.5 and $15\,\mu$m ($4000-667\,\mathrm{cm}^{-1}$), known as midinfrared, and this portion of the spectrum has been traditionally used for structural identification studies.

Modern analytical instrumentation, based on infrared measurements, can determine nondestructively the composition of some foods, such as milk, dairy products, and grains. Although when separated each of the food constituents, water, protein, fat, or carbohydrate, has a unique optical absorption fingerprint in the infrared, when an integral food sample is analyzed many mutual interferences occur, i.e., at the same wavelength more than one constituent absorbs radiation. For example, the absorptions of hydroxyl groups of water and sugars, and of alkyl and carbonyl groups

Analytical Applications of Infrared Radiation

of proteins and fats, overlap. Therefore, for each food constituent, the measurements are made at at least two wavelengths, one at the maximal and one at the minimal absorption. These instruments measure accurately and simultaneously the minute light absorptions or reflections at several wavelengths and discriminate quantitatively between constituents using digital-processing equipment with proper, memorized, calibration data.

One of the most successful analytical applications of quantitative infrared analysis is that of the concomitant determination of protein, fat, lactose, and water in milk. The fat concentration is quantified either by the number of ester linkages in triglycerides, by monitoring the carbonyl group absorptions at $5.73\,\mu m$, or by the number of carbon–hydrogen bonds in the fatty acid chains, monitored at $3.45\,\mu m$. The latter absorption, in addition to the fat, measures all carbon–hydrogen bonds in the sample, including those in protein and lactose. At the infrared absorption wavelength for protein determination ($6.4\,\mu m$), it is the nitrogen–hydrogen bond within the peptide link which absorbs the radiation, and the extent of absorption is a function of the number of amino acids present. Finally, the absorption of the bond between the hydroxyl group and the carbon atom at $9.55\,\mu m$ is monitored for lactose concentration. The mathematical treatment of the data individualizes the component concentrations. The concentration of the total solids is not measured directly, like the other constituents, but is computed by summing up the concentrations of fat, protein, lactose and a constant, arbitrarily chosen, value for mineral content.

Although the highest vibrational frequencies of molecules correspond to light of wavelengths near $2.5\,\mu m$, molecules also absorb light at shorter wavelengths in the region defined as "near infrared" from 0.8 to $2.5\,\mu m$ ($12\,000–400\,cm^{-1}$). A molecule can absorb infrared radiation when the frequency of the light is an even multiple of the original vibrational frequency. These absorptions are called "overtone bands" and they become weaker as the multiple becomes larger. The first overtone bands are at a frequency of twice the vibrational frequency, the second overtone bands are at three times the vibrational frequency, and so on. This wavelength region between the visible and $2\,\mu m$, near infrared region, offers important advantages for food analysis (Osborne and Fearn 1986; William and Norris 1987). The infrared spectra are simpler in the region of weaker overtone bands and the bands of protein, oil, sugar, and moisture are still strong enough to be measured accurately. This is particularly relevant for water which absorbs very strongly through most of the infrared, and it is usually preferred that samples are dried for midinfrared analysis. This is not necessary in the near infrared, where absorption bands of water are strong enough to allow moisture content to be measured, but are not so strong as to give problems in making other measurements. Furthermore, glass does not transmit wavelengths longer than $2.5\,\mu m$ and cannot be used for sample-holding cells in the mid-infrared. Conversely, there is no restriction for the use of glass in the near infrared measurements. Since 1970 there have been major advances in the generation of near infrared spectra, in electronics for detection of low levels of light, and in low-cost computation capability; these have eliminated the prior limitations of near infrared analysis.

Based on the same principles as those as absorption infrared spectroscopy, but requiring different instrumentation, near infrared reflectance spectroscopy provides rapid and precise measurements of protein, oil, moisture, fiber, and other key organic components in opaque samples like grains, feeds and other solid foods (Finney 1978). In summary, quantitative infrared analysis allows the almost instantaneous measurement of the chemical composition of a food sample without requiring special sample preparation or destruction of the sample.

Infrared CO_2 analyzers have been suggested for use in pest management programs. They can detect a single larva of *Ephestia cautella* Wlk in 10 lb of dates and can detect hidden infestations in wheat grain (Street 1976).

A different application for infrared radiation in food analysis is the heating lamps which are extensively used for fast drying of samples in the determination of total dry matter.

Chapter 4   Microwave Radiation

4.1 The Definition of Microwave Radiation

In the upper boundary of the frequency scale of electromagnetic radiation are the radioelectric signals. The frequency range extending from 300 GHz (3×10^{11} Hz or wavelength of 1 mm) down to 300 MHz (3×10^8 Hz or wavelength of 100 cm), which characterizes signals having between 300 million and 300 billion periods per second (the period of a signal is defined as the inverse of the frequency, and is located between 3 ns and 3 ps), is generally known as microwave radiation. These limits are to some extent arbitrary but they permit position of the microwave domain between infrared rays, at the higher end of the frequency range, and radio and TV waves at the lower end. (The prefix *micro* may lead to the expectation of wavelengths in the micrometer range and not, as is actually the case, in the millimeter to meter range). Even weak chemical bonds, such as hydrogen bonds, have energies that are several orders of magnitude greater than the energy of microwave photons. Not until the frequency of a radiation approaches, or exceeds, 10^{16} Hz do photon energies become comparable to the binding energy of electrons to atoms. It seems quite resonable, therefore, that microwave photons cannot disrupt the electronic structure of atoms, nor can they disrupt chemical bonds.

Microwaves are mostly employed in three kinds of applications: radiocommunications, radar, and heating. Naturally, the last one is of immediate relevance to the food industry.

4.2 Molecular Mechanisms of Heating with Microwaves

As with infrared heating, microwaves heat by radiation. Nevertheless, quantitatively, these two methods differ significantly. While in infrared heating, the penetration of the radiation in the target is minimal and most of the food mass is heated by conduction, with microwaves penetration into the food is much deeper, reducing the heat diffusion distance and time, so that the principle of radiation or "volume heating" is better accomplished.

Microwaves, as with any electromagnetic radiation, are emitted or absorbed only in photons of discrete quantities. For a molecule to absorb a photon, the energy of the photon must exactly match the energy difference

116 Microwave Radiation

between the ground state and some other allowed energy state. This process is usually described for microwave as "coupling" or energy transfer between the radiation field and the molecule. The coupling of the microwave electromagnetic field to molecular systems, with heating as the final result, takes place through the force that the electric field component of the radiation exerts on charged parts of the absorbing molecules. Magnetic field interactions would be important only if the systems involved were ferromagnetic or paramagnetic, which is not the case for food materials. Therefore, all magnetic effects can be ignored.

Microwaves are either reflected, transmitted, or absorbed, depending upon the material contacted. In materials with conduction electrons, such as metals, the free electrons are moved by the electric field of the incident microwave radiation. The movement is driven by forces directed to annihilate a radiation-induced electric field in the metal. The resulting changes in the metal cause the reflection of the incident wave and, consequently, microwaves cannot penetrate metals. In theory, at least, metals are not heated by the microwave field. In practice, some metals, being imperfect electric conductors, can exhibit some degree of heating.

Conversely, microwaves can be transmitted through many materials without being absorbed. Perfect electric insulators transmit microwaves without absorbing energy. Air is virtually transparent to microwaves and since no interaction is taking place, microwave propagation is unobstructed. Materials such as glass and most plastics can be used to contain food within a microwave applicator without obstructing the flow of energy into the food, and therefore without being heated.

In between the perfect electrical conductors and insulators are the composite materials, like foods, which absorb microwaves. In foods the reflection of microwaves is much less significant than absorption and transmission, although at any interface between materials with different electric properties a partial reflection will occur.

The material subjected to microwave radiation is often referred to as the "load". The absorbed microwave energy is manifested as heat upon interaction with the load as a result of several complex energy transfer mechanisms. These processes, which are generally described as "molecular frictions", involve dipole rotations, stretching of polar molecules, conductive migration of dissolved ions, and induced polarizations.

When an external alternating electromagnetic field is applied the load undergoes electric polarization in the process (Fig. 22). As the wave travels by an absorbing medium, the molecules will experience the oscillations of the electric field component. Simplistically, if the axis of a molecular dipole is perpendicular to the electric field, torque forces exerted on the molecule will tend to set it into rotation. If, however, the wave oscillates along the axis of a dipolar molecule, it tends to stretch and compress it, setting up vibrations. This transfer of energy from field to molecule occurs only if the

Molecular Mechanisms of Heating with Microwaves

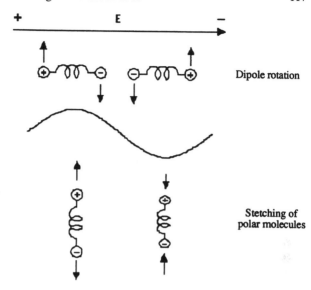

Fig. 22. Illustration of forces exerted on polar molecules by the electric field of an electromagnetic wave. (After Curnutte 1980)

frequency, or energy, of the photon matches one of the energy differences in the allowed rotational or vibrational levels of the medium. The molecular motions are synchronized with the oscillating electrical field. If the energy absorbed from the field is to be converted into heat, these internal molecular motions are to be conveyed to the surroundings (Curnutte 1980).

Water, which accounts for much of the primary absorption of microwaves in biological materials, and therefore in foods, is a polar molecule. Its total electric charge is distributed in such a way that one end of the molecule is positive and the other end is negative. When placed in an electric field, positive and negative charges move in opposite directions and the water molecules will tend to rotate so that their dipoles align with the field. This degree of alignment is referred to as the polarization of the sample. The rotation rate of water molecules in the liquid medium is determined by the forces of friction which result from the interaction with neighboring molecules in the medium. Whenever the electric field does work in moving electric dipoles, energy is absorbed and converted to heat. No work is performed by a field which is alternating at low frequencies, since when the electric field changes direction, the water molecules are rotating so rapidly that they reorient themselves without any help from the field. On the other hand, at very high frequencies, none of the water molecules in the load are able to follow the field and, again, no work can be performed. As the radiation frequency exceeds 20 GHz less energy is absorbed and eventually the absorption decreases to zero at about 1000 GHz.

If the water molecules are well separated, as in the gaseous phase, then there is only a relatively small number of allowed rotational and vibrational transitions, and only photons of very distinct frequencies will be absorbed. Microwaves of any other frequencies will be transmitted without energy loss. On the other hand, in a liquid medium, hydrogen bonding brings the water molecules in very close contact and causes the allowed transition levels to become nearly continuous and to cover wide bands of microwave frequencies. Therefore, almost any incident microwave frequency within these bands matches some allowed transition; it interacts with the charges involved in hydrogen bonding of water molecules and is absorbed.

There is little doubt that the interaction of foods with microwave radiation is dominated by water. In addition, however, the inorganic ions of salts dissolved in food also interact with the electric field by migrating toward oppositely charged regions of the field. As a result, the water hydrogen bonds are disrupted with generation of additional heat.

Nonpolar molecules can also be polarized. In the normal state, the opposite charges in such molecules are symmetrically balanced. When an electric field is applied, the positive charges experience a force in the direction of the field and the negative charges are repelled opposite to the field direction. This induced polarization then allows the electric field of the electromagnetic wave to interact with the molecule in a fashion similar to that experienced by dipoles. This mechanism explains the energy transfer from a microwave field to molecules of oils and fatty tissues.

The process of heating through permanent and induced polarizations is called dielectric heating. The overall process of heating by microwaves is defined by the dissipation of electric energy in "lossy" media (the word "loss" refers to this process of dissipation of electric energy within dielectric materials). This feature is quantitized by parameters like dielectric loss, loss tangent, or loss factor.

The ability to polarize, namely the orientation of dipoles and their moments induced by a field, determines the dielectric constant within the medium. The dielectric constant of a medium, ε', is defined by the following equation:

$$\varepsilon' = \frac{QQ'}{Fr^2}$$

where F is the force of attraction between two charges, Q and Q', separated by a distance r in a uniform medium. The relative dielectric constant is defined as:

$$\varepsilon'_r = \frac{\varepsilon'}{\varepsilon_0}$$

where ε_0 is the dielectric constant of air.

The polarization effect is a function of the radiation frequency, the dielectric and electric properties of the material, the viscosity of the medium, and the size of the polar molecules.

The dielectric relaxation is the time, usually a fraction of a microsecond, during which the molecules delay in responding to the field change, or, conversely, it is the time to revert to disorientation. In the process of relaxation is found the basis for dielectric loss. The dielectric loss implies the total effect of energy dissipations through elastic distortions, deformations, and displacements which occur under field stimulation and relaxation forces. Since the relaxation time depends on the viscosity of the medium which, in turn, depends on the temperature, the loss will therefore be temperature dependent.

As the radiation frequency increases, the dielectric constant remains the same or decreases, but the absorption passes through a maximum which is related to the relaxation time. Thus, the maximum absorption occurs when the angular frequency, ω ($\omega = 2\pi v$), in radians per second, equals the reciprocal of the relaxation time, τ:

$$\omega = \frac{1}{\tau}$$

The frequency of maximum absorption is called the resonance frequency and is specific for a certain dipole, its field and medium, all of which combine to define a relaxation time. With small molecule polarizations in low viscosity media the relaxation time would be 10^{-11}–10^{-12} s and the maximum absorption would be in the range of microwave frequencies. For example, the corresponding wavelength for the maximum absorption of water ranges from 0.61 to 3.37 cm as the temperature decreases from 75 to 0 °C. If adequate power is available at the correct frequency, the process of absorption is further characterized by a very dramatic rate of heating throughout the volume of absorption.

When the absorptive polarization occurs, the total current can be resolved into two current components, the polarizing and the loss currents, which are 90° out of phase. The dielectric constant (ε') and dielectric loss factor (ε'') reflect the energy transfer from the field to the material and are measures of the microwave absorption. The dielectric constant reflects the ability of a material to store electromagnetic energy and the dielectric loss factor represents the ability to dissipate it, that is, to convert it into heat. Both parameters can be represented as Cartesian vectors with the vector sum as the aggregate effect (Fig. 23; von Hippel 1954).

The resulting vector is the material's complex permittivity, which is related to its ability to interact with an electric field and dissipate field energy as heat. The angle δ between the polarizing and total currents is the loss angle. The tangent of the angle δ, also called the dissipation factor, represents the energy loss characteristic of the material:

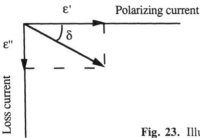

Fig. 23. Illustration of a material's complex permittivity

$$\tan \delta = \frac{\varepsilon''}{\varepsilon'}.$$

Tan δ is of special interest in microwave heating since it reaches a maximum value at the frequency of maximum absorption.

The energy transported by microwaves is measured in units of electric power, P, (watts) according to the following relationship:

P = VI

where V is the current potential in volts and I is the current intensity in amperes. Since

V = IR

where R is the resistance in ohms,

P = I²R

which is defined as ohmic loss or the total electric energy available for conversion into heat.

The average microwave power absorbed, which can be converted into heat for a unit volume, P_{abs} (W/cm³), is given by:

$$P_{abs} = 2\pi \nu E^2 \varepsilon_r''$$

where E is the electric field strength in volts per unit distance.

The rate of temperature increase, dT/dt, is given by:

$$\frac{dT}{dt} = \frac{k\nu E^2 \varepsilon_r''}{\rho C_p}$$

where k is a constant, ρ is the density of the material in g/cm³, and Cp is the specific heat in cal/g °C.

Inside the absorbing material heat is generated in a distributed manner allowing, in theory, for a uniform and fast heating. Although in contradistinction to convection or conduction, microwave heating per se is a "volume" heating; in reality the overall heating effect is a combination of "volume" and thermal conduction. The relative contribution of these heat-

Molecular Mechanisms of Heating with Microwaves

ing processes is initially dependent upon the depth of penetration of the microwaves. The penetration can be infinite in perfectly transparent media, such as loss – free glasses and ceramics, zero in reflective materials, like metals, or can have definite values in foods. The depth of penetration is defined as the distance from the surface to that at which the incident power drops to 37% (1/e).

The attenuation of the incident radiation power (P_0) to the penetration depth, d, is exponential:

$$P = P_0 e^{-2\alpha d}$$

where α is the attenuation constant. The α value of a medium is defined as follows (Copson 1975):

$$\alpha = \frac{\pi \varepsilon''}{\lambda_0 \sqrt{\varepsilon_r'}}$$

where λ_0 is the radiation wavelength in free space.

It follows that attenuation relates directly to the radiation frequency or, reciprocally, the penetration in the microwave range increases as the frequency decreases. As expected, attenuation is also a function of material composition as reflected by dielectric properties.

Penetration is also expressed by the depth at which half of the incident energy is absorbed, D_{50}, which is (von Hippel 1954):

$$D_{50} = \frac{\lambda_0}{2\pi} \sqrt{\frac{2}{\varepsilon'(\sqrt{1 + \tan \delta^2} - 1)}}$$

Details of penetration and absorption become much more complicated when the load is nonhomogeneous. The problem arises because reflections occur whenever microwaves strike the interface separating two different materials. The reflected waves from the interface of two layers return to the previous interface and are transmitted and/or reflected again. These reflected and transmitted waves interfere with one another and, depending on the thickness of the layer, can cause cancellation or reinforcement of the wave. These phenomena create a potential for preferential heating of certain layers.

In practice, heating by microwaves is controlled by a multitude of interwoven factors, some related to the radiation source, like frequency and output power, and others, like dielectric properties, electric conductivity, heat capacity, and thermal conductivity, related to the properties of the food exposed to radiation. These latter properties are dependent on factors of the chemical composition, such as moisture, salt, sugar, and fat contents, and on physical parameters, such as bulk density, mass and even physical geometry. These factors and their relevance in food processing have been summarized by Schiffmann (1986).

Dielectric Properties. A summary of the dielectric properties of various food and related materials is given in Table 16.

In the conversion of dielectric loss to heat, the frequency response of the dominant component will dictate the overall heating rate. The dielectric properties of foods at microwave frequencies are, for many practical pur-

Table 16. Dielectric properties of various food and related materials

Material	Dielectric constant (ε')	Dissipation factor ($\tan \delta$)	Dielectric loss factor (ε'')	Penetration (D_{50}, cm)
Water (25°C)[a]	76.7	0.157	12	1.25
Water (1.5°C)[a]	80.5	0.31	25	0.7
Ice (-12°C)[a]	3.2	0.0009	0.0029	700
0.1 M NaCl (25°C)[a]	75.5	0.24	18	0.8
Water vapor (at low pressure)[a]	1		0	
Air, carbon dioxide[b]	1			
Meat steak (25°C)[a]	40.0	0.3	12	
Dried meat (25°C)[a]			0.00096	
Raw meat (25°C)[c,d]	49	0.33	15	
Cooked meat (25°C)[c,d]	45	0.56	12	
Beef (25°C, 915 MHz)[b]	55		22	
Beef (65°C, 915 MHz)[b]	51		33	
Beef (25°C, 2450 MHz)[b]	52		17	
Beef (65°C, 2450 MHz)[b]	48		18	
Pork (25°C, 915 MHz)[b]	54		23	
Pork (65°C, 915 MHz)[b]	52		34	
Pork (25°C, 2450 MHz)[b]	51		17	
Pork (65°C, 2450 MHz)[b]	49		20	
Ham (20°C, 2800 MHz)[e]	54		28	
Ham (60°C, 2800 MHz)[e]	64		41	
Cooked carrots (25°C)[c]			17	
Mashed potatoes (25°C)[c,d]	65	0.34	22	
Potato (20°C, 900 MHz)[f]	68		20	
Potato (60°C, 900 MHz)[f]	59		26	
Potato (25°C, 3000 MHz)[f]	66		19	
Potato (65°C, 3000 MHz)[f]	64		17	
Carrot (20°C, 2800 MHz)[e]	72		18	
Carrot (60°C, 2800 MHz)[e]	61		15	
Paper[d]	2–3	0.05–0.1		
Pyrex glass (25°C)[g]	4.5–6.0	0.0054	0.025–0.0081	
Porcelain[g]	6.0–8.0		0.01–0.003	

[a] From Copson (1975).
[b] From To et al. (1974).
[c] From Risman and Bengtsson (1971).
[d] From Carroll (1989).
[e] From Bengtsson and Risman (1971).
[f] From Ohlsson et al. (1974).
[g] From Weast (1967).

poses, determined by their water and salt contents . Water is the major absorber microwaves in foods and, consequently, the higher the moisture content, the better the heating. The organic constituents of food are dielectrically inert ($\varepsilon' < 3$ and $\varepsilon'' < 0.1$) and, compared to aqueous ionic fluids or water, may be considered transparent to microwaves. Only at very low moisture levels, when the remaining traces of water are bound and unaffected by the rapidly alternating microwave field, do the components of low specific heat become the major factors in heating. In high carbohydrate foods, such as bakery products, syrups, and alcoholic beverages the dissolved sugars and alcohol are the main microwave susceptors. The presence of bone in meat also affects the uniformity of microwave heating because calcium and other minerals in bones reflect the microwaves which penetrate the muscle tissue; heating is therefore greater at the contact surface with the bone than in other areas of the meat.

Foods with phases of diametrically opposed dielectric properties are likely to be heated with drastically different temperature gradients: highly absorbing components suspended in a continuous phase of low absorbance (partially defrosted areas in frozen foods or the jelly in doughnuts), low absorbance components suspended in a continuous phase of high absorbance (meat and vegetable pieces in soups), or layered products with alternating phases of low and high absorbance (cheese and dough layers in pizzas) (Mudgett 1986). Occasionally, the very uneven heating could be beneficial, as in a microwavable sundae, which is an ice cream novelty with a warm topping over cold ice cream.

All materials, other than reflective metal conductors and entirely loss – free transparent substances, will heat eventually in a microwave oven. However, if a material of high loss is simultaneously present with a material of low loss, the latter may never be heated.

Intrinsically, the dielectric constant of a compound is a parameter which is dependent on temperature. This property is strikingly evident in the phase transition of ice to water. While the dielectric properties of water are $\varepsilon' = 80$ at 20 °C, $\tan \delta = 0.16$ and $\varepsilon'' = 12.8$, which characterize a good absorber of microwaves, the values for ice of $\varepsilon' = 3.2$, $\tan \delta = 0.0009$ and $\varepsilon'' = 0.0029$ are representative for a highly transparent compound. These diametrically opposed behaviors explain the "runaway" problem, wherein frozen food thawed at one point attracted a disproportionate share of the energy and was heated excessively while the remainder was still frozen solid.

At a frequency of 915 MHz, the loss factors for raw potato (Mudgett et al. 1977) or beef (To et al. 1974) increase with temperature. At 2450 MHz frequency, however, the loss factor slightly decreases to a minimum near 40 °C, and then increases at higher temperatures. This behavior was related to the opposite effects of temperature on dipole rotation of water molecules and conductive migration of dissolved ions. That is, on the one hand, the dipole loss of the water decreases as a result of temperature effects at the critical wavelength and, on the other hand, the ionic loss increases as a result of temperature effects on equivalent conductivity (Mudgett 1986).

In addition, the electromagnetic properties of foods may vary considerably during the heating process if the moisture level in a multicomponent food is allowed to decrease with the increase in temperature. In such a case, the heating process slows down with the evaporation of water from the food and the cooking process becomes then self-limiting. Fatty foods exhibit an opposite dependence.

Electric Conductivity. In addition to dipolar rotation, ionic conductance is a major molecular mechanism for generating heat under microwave exposure. In terms of food, this property is mainly controlled by the amount of salt present. The addition of salt, which can increase the heating rate and affect the depth of penetration, is a factor to be taken into account in food formulations for microwave processing.

Radiation Frequency. Only certain bands are allowed by law for industrial, scientific, or medical use, because of the need to avoid interference with bands used for communication. The frequency bands currently used for microwave heating are 2450 ± 50 MHz, covering the frequency used in domestic ovens (2450 MHz, central wavelength 12.2 cm) and 970 ± 25 MHz (central wavelength 32.8 cm). Other bands reserved for these uses are 5800 ± 75 MHz (central wavelength 5.2 cm) and $22\,125 \pm 125$ MHz (central wavelength 1.4 cm).

The penetration depth of the incident energy is inversely related to the frequency, though this effect can be abrogated by other factors such as salt and water contents. For example, the D_{50} in water for radiations of 2450 and 915 MHz at 40 °C are approximately 3 and 20 cm respectively, but for gravy at the same temperature penetration depth is around 7 mm at both frequencies.

The dielectric properties of a material are also intrinsically affected by frequency. For example, for pure water at room temperature the loss factor (ε'') is approximately three times as large at 2450 MHz than at 915 MHz, while for $0.1\,M$ NaCl solution the loss factor at 915 MHz is only more than twice as large.

On appearance, it might seem advantageous to select the microwave frequency by an inverse relationship with the size of the material subjected to heating. In practice, however, the radiation frequency effect is often of less importance in foods as it is attenuated by the complexity of chemical compositions and other heat mechanisms.

Several other factors indirectly affect the overall efficiency of heating by microwaves.

Product Bulk Density. Foods can exist in a looser or more compact state depending on the air content, which means that for the same chemical

Molecular Mechanisms of Heating with Microwaves 125

composition the bulk density varies. Foods of a lower density heat faster at a given level of power absorption than do denser foods of similar composition. Air is practically transparent to microwaves because of the low dielectric constant ($\varepsilon' = 1.0$) and therefore its inclusion in a material reduces the amount of microwave power absorbed and increases the penetration depth. While conventional heating by conduction of airy foods, like bread dough, is slowed down by the thermal isolation provided by the inclusion of air, microwave heating of such foodstuff is even faster because of the deeper penetration.

Specific Heat. The thermal capacity of a substance, Cp, is the quantity of H calories spent to raise the temperature of m grams of a substance from temperature t_1 to t_2 in °C:

$$C_p = \frac{H}{m(t_2 - t_1)}.$$

The specific heat of a substance is the ratio of its thermal capacity to that of water at 15 °C.

In spite of a low dielectric loss and a poor microwave absorption, a compound of a low specific heat can still be heated up easily by exposure to microwaves because of the lower amount of calories needed per unit weight to raise the temperature. In the context of foods this property is most relevant for fats and oils. Although the relative dielectric constant ($\varepsilon' < 3$) and loss factor ($\varepsilon'' < 0.1$) (Pace et al. 1968) of oils and fats are much smaller than those of water ($\varepsilon' = 80$ at 20 °C), they heat considerably faster since the thermal capacity of 2 kJ/kg °C is less than half that of water (4.2 kJ/kg °C).

Thermal Conductivity. Roughly speaking, while traditional techniques heat a surface, microwaves heat the whole volume of the material treated. Since, however, the heating depth of microwaves is limited by the extent of penetration, the thermal conductivity of the material may influence the heating homogeneity. This is particularly true for long heating times and for large amounts of food.

In microwave heating the food surface which is in contact with cold air remains cooler than in conventional heating, and less heat is lost to the environment. If desired, the superficial heating may be enhanced by insulating the target food from the surrounding air by wrapping, dipping in lossy material (gravies), etc.

Mass. In any method of heating there is a direct relationship between the amount of energy available for heating and the total amount of material or sizes of individual pieces subjected to heating. Since in microwave heating the primary and immediate factor is the depth of penetration and thermal conduction is second to it, this relationship is even more stringent. If the

126 Microwave Radiation

size of an individual piece is very large in comparison to the wavelength, superficial heating is favored, whereas for sizes closer to the wavelength, temperature may be highest in the center. It also follows that the more regular the shape, the more uniform the heating. Thinner parts may be overheated compared to larger parts. This effect may be controlled by reducing the microwave power input and extending the heating time.

Microwave Power Output. For microwave installations power outputs range from a few hundred watts in domestic ovens to tens of kilowatts in industrial systems. Although the speed of heating can be easily increased by boosting the microwave power, in practice, this option is treated with caution. An excessive rate of heating leads to a nonuniform profile of temperature distribution since the transfer of heat within the food lags behind local rises in temperature. In general, the rate of microwave heating is related to the power output of the heating installation more significantly than in conventional heating. While in conventional heating the limiting factor of the heating rate is the thermal conductivity of the material, with microwaves the heterogeneous dielectric properties are the determinant. Each processing operation, such as cooking, proofing, baking, and drying, requires specific optimized heating gradients to enable the complex milieu of desired physicochemical changes to occur properly. Nevertheless, some recipies may command jiffy microwave heating. Thus, the food poisoning risk posed by egg-based sauces can be reduced by a combination of acidification and rapid boiling, which reliably eliminates salmonellae from yolks without curdling them or fatally compromising their ability to produce smooth, stable emulsions (McGee 1990).

4.3 Equipment for Microwave Heating

Microwaves for heating are generated by an electronic device known as the magnetron. Magnetrons, operating most often at the 2450 MHz frequency, generate microwave powers between a few hundred watts and a few kilowatts, depending upon the application. The electromagnetic energy is conducted from the magnetron to the site of use by a wave guide (Decareau and Peterson 1986). For 2450 MHz frequency, the wave guide can be a rectangular metallic tube of 3.75 × 2.15 in. cross section with 0.008 in. walls.

The device used to transfer the microwave energy to a load is called the applicator and these may be classified in one of four categories:

1. A metal enclosure which is actually a resonant cavity and serves as an oven. This kind of applicator is most commonly used in the food industry. The release of microwaves in a small container create specific technical

Equipment for Microwave Heating

problems. When a microwave beam is directed toward a sheet of metal, such as the oven wall, then the beam is reflected. In the region near the wall the reflected beam will meet the direct wave from the source traveling in the opposite direction. Therefore, the fields of reflected and incident waves cancel at the wall surface and at each integral number of half wavelengths from the metal sheet. In between these positions, called nodes, the electric field oscillates with twice the amplitude it would have if only one wave was present. This combination of two oppositely traveling waves is called a standing wave. These standing waves affect the distribution of energy at the surface of the load and may lead to uneven heating within the product because of variations in intensities of surface power per unit area. This problem is basically related to the oven design and can be minimized by mode stirring. A mode stirrer, which is a kind of fan with metallic vanes, distributes the microwave energy among the different resonant modes of the cavity ensuring homogeneous heating.

By a similar principle, this phenomenon may occur within the food. There is a possibility, more likely for materials with a very low loss factor, of a standing wave pattern within the product, caused by interaction between internally transmitted waves and those reflected from the opposite surface. Such internal reflections create internal hot spots and may help to explain the heating of oil and other materials with low values of dielectric constant and loss. Internal standing waves seem unlikely for high- or intermediate-moisture compounds with relatively high attenuation factors and low field penetration depths.

The electromagnetic properties of foods vary considerably during the heating process. Because of nonlinear temperature changes, the effect of radiation absorbers in the oven, and the difficulty in accounting for the inhomogeneity of the radiation flux in a microwave oven, it is practically impossible to maintain constant parameters. The changes in the loss factor with temperature cause the power actually absorbed by a food material to be temperature-dependent. Consequently, the load cavity and the generator resonant circuit may become unmatched. Cavities for heating are designed for a preferred reflection from the walls toward the load rather than toward the power tube. This makes them more resistant to changes so that a rated power output can be maintained and the equipment is able to withstand large mismatches.

The effects of large differences in microwave absorption between various regions of the food are minimized by permitting heat conduction to equalize the temperature differences. Therefore, the ovens are built with the capability for pulsing on and off the energy input. This allows the inward conduction of heat in the food and temperature equilibration.

The temperature measurements in microwave ovens cannot be accomplished with metallic thermometers or thermocouple probes. Measurement equipment made of loss-free materials should be used. In domestic microwave ovens probes comprising high resistance conductors and tiny thermistors are used. This works reasonably well as the tip is immersed in the food and is shielded from the microwaves, but it becomes useless if the shielding effect is lost, like during drying operations. In sophisticated equipment, infrared devices, which use the emissions from the surface connected to thermal profile measurement systems, like thermal video systems, and fluoroptic thermometry, are employed. Infrared thermography offers speed, accuracy, and resolution in measurements of temperature and records of heating patterns of cooked foods. The fluoroptic method is based on the measurement of intensity of two sharp fluorescent emission lines from an europium-activated phosphor, which show dissimilar variation in intensity with temperature. The sensor is made of phosphor powder bonded to the tip of a fiber optic probe which is temperature resistant and transparent to microwaves. Ultraviolet light transmitted down the fiber excites the phosphor and the fluorescences are compared with a standard calibration curve.

2. Slow wave applicator. Some industrial ovens are designed for continuous use. In these, the material which is allowed to move, for example, on a conveyor belt through the cavity close to an open transmission line, interacts in a continuous fashion with the fringing electromagnetic field of the line. The end sections of the cavity are designed to prevent the microwaves from leaking toward the outside. This field is quite inhomogeneous, but as the material flows, the average heat may be maintained at an approximately constant level.

3. Traveling wave applicator. This kind of applicator is primarily used for sheet materials or for textile threads which cross a slotted rectangular waveguide.

4. Free space applicator. This kind of applicator is an antenna designed to irradiate rather bulky elements which cannot be fitted within an enclosure and which can create very large thermal gradients. At lower power levels such applicators are used for local heating in medical treatments.

There is currently no simple method that permits true dosimetry on target food exposed to microwave radiation. The power incident to the food surface, expressed as power density (mW/cm^2), is the most common parameter used to approximate the energy absorbed by the target. This measurement is usually conducted by means of a standard-gain receiving antenna positioned in the far-field, or by calculating the power density from a knowledge of the power delivered to a standard-gain transmitting antenna and the distance from the antenna. The power density contours are changed when an object is inserted into the field, and this, coupled with reflection of the beam, introduces uncertainties into these methods of estimating the incident power. Likewise, the energy absorbed can be approximated by calorimetry conducted at the position of the target food, but the problems

Applications of Microwave Heating in the Food Industry 129

inherent in the assumptions required with this technique mean that the measurements are not very accurate.

4.4 Applications of Microwave Heating in the Food Industry

In 1945, a manufacturer of radar magnetrons found, supposedly by chance, that microwaves can produce heating (Spencer 1945). This effect was promptly utilized in the design of ovens for heating food in commercial establishments, and for processing raw food materials in industry. Since the beginning of the 1970s, microwave ovens have appeared increasingly in family kitchens and have become the predominant household appliance based on microwave technology.

The most remarkable property of microwaves, that of in-depth heating, has been considered for a large number of applications in food processing. Gerling (1986) counted hundreds of patents and other publications for the period 1969–1984 that involved the use of microwaves in the following types of food processes: baking, blanching, browning, coagulation, coating, compacting cooking, curing, drying, fermenting, freeze drying, gelatinization, heating, making, pasteurization, preservation, processing, proofing, puffing, rendering, roasting, separation, shucking, sterilization, tempering, thawing, and vacuum drying. Table 17 summarizes some of these applications (Decareau 1986; Decareau and Peterson 1986).

In spite of this impressive list, to-date there are still relatively few industrial applications of microwave heating, most probably because of limiting technical and economic factors. Tempering frozen foods, drying pasta and onion, and cooking bacon account for the majority of industrial microwave installations. Many of these processes combine microwave and conventional heating methods (Decareau 1986; Mudgett 1989). As such,

Table 17. Applications of microwaves in food-processing operations

Operation	Products
Tempering	Meat, fish, poultry, butter
Cooking	Bacon, meat, patties, sausage, potatoes, sardines, chicken
Drying	Pasta, onions, rice cakes, egg yolk, snack foods, seaweed
Vacuum drying	Orange juice, grains, seeds
Freeze drying	Meat, vegetables, fruits
Pasteurization	Bread, yogurt, milk
Sterilization	Pouch-packed foods
Baking	Bread, doughnuts
Roasting	Nuts, cocoa beans, coffee beans
Blanching	Corn, potatoes, fruit
Rendering	Lard, tallow

130 Microwave Radiation

the industrial implementations are still relatively modest as compared with the phenomenal growth of the domestic microwave oven market. Introduced as a luxury item in 1967, microwave ovens have became a standard household appliance, due to their great convenience, flexibility and time saving. The increase in working households has boosted the demand for meals that can be rapidly prepared. Partially precooked and prebrowned dishes have become standard items for the consumer who has less time for meal preparation and frequently has to plan around fragmented mealtimes.

Microwave heating possesses several potential advantages:

1. Convenient and versatile source of heat, quite flexible and instantaneously responsive to control. Most heating can be accomplished faster than conventional heating.
2. Clean.
3. Heat processing of packaged products is possible in thermal isolator packs.
4. Heating is quite effective since the surrounding air and oven are not heated up. The portion of electric energy drawn from the network that is transformed into heat reaches 40–50% for microwave ovens operating at 2450 MHz. The losses are evaluated as 5% in high voltage supply and driver, 40% in magnetron, and 10% in waveguide and applicator.
5. Savings in floorspace since the same throughput of food can be achieved with equipment that is smaller than the conventional equipment.
6. Shorter heating cycles mean faster processing which result in savings in storage space.
7. Microwave heating is unique in that it can advantageously be used in combination with other heat-transfer sources to obtain a desirable result (surface air-cooling in microwave thawing of frozen foods or browning by conventional heating in microwave baking).
8. There is a reduction of in-process shrinkage and losses.
9. Since microwaves heat internally, temperature distribution may be more uniform and overheating of the surface can be avoided. On the other hand, since the absorption of the microwaves is dependent on the electromagnetic properties, some food components may heat faster than others. This enables unusual temperature profiles in multicomponent foods and could be beneficially used for composite dishes.

In contrast to these advantages there are the following drawbacks:

1. Since the absorption of microwaves is dependent on the electromagnetic properties, the temperature profile in a multicomponent food could be highly erratic. Uneven heating can occur due to product characteristics, size, and shape. Overheating tends to occur on sharp corners and edges. Underheating can occur in the center of large pieces of food.

Applications of Microwave Heating in the Food Industry 131

2. Capital costs are high; magnetrons are more expensive than conventional heating elements.
3. More complex technology is involved, resulting in maintenance difficulties for unskilled personnel.
4. Microwaves cannot grill, brown, or crust food surfaces.
5. Safety requirements are different from those of conventional heating.

In theory, microwaves can be used in any processing unit involving the application of heat. Specific applications of microwave heating in food technology have been reviewed by Rosenberg and Bohe (1987a,b).

Pasteurization and Sterilization (effects on microorganisms). The killing of pathogens and diminution of populations of spoilage microorganisms in food are the major goals of pasteurization. The sterilization of a food product requires the deactivation of the entire population of microorganisms, either in the vegetative state or as spores.

In the wider context of interactions between microwaves and biological entities, the effect of microwave irradiation on microorganisms in model experimental systems has been a topic of extensive research (Chipley 1980; Fung and Cunningham 1980). In principle, there should be no difference between the resistance of various microorganisms to microwave or conventional heating since it depends on the same time–temperature relationships. However, in practice, it is difficult to accurately compare there under "identical experimental conditions". When both techniques were compared in terms of final condition and temperature of the product, conventional cooking and heating generally yielded lower colony counts by one or two orders of magnitude (Bengtsson et al. 1970; Blanco and Dawson 1974; Crespo and Ockerman 1977; Crespo et al. 1977; Ockermann et al. 1976).

High-temperature short-time (HTST) and low-temperature long-time (LTLT) pasteurization conditions were simulated using uninoculated and inoculated milks that were heated in microwave ovens. Milk samples heated for 59 s at 700 W achieved a temperature of 71.7 °C, prescribed for HTST pasteurization, but even heating for 65 s and maintaining this temperature for an extra 15 s failed to inactivate all the added cells of *Salmonella typhimurium*, *Escherichia coli*, or *Pseudomonas fluorescens*. Alternatively, milk was heated to a temperature between 62.8 and 71.7 °C for 4.5–5 min, depending on the power that was used (550 or 700 W), and was then refrigerated overnight. Such treatments also failed to reduce the population of *Streptococcus faecalis* in the milk by the degree that occurred when inoculated milk was heated in a water bath at 62.8 °C for 30 min (Knutson et al. 1988). Cooking whole turkeys inoculated with three common pathogenic foodborne bacteria (*Salmonella typhimurium*, *Staphylococcus aureus*, and *Clostridium perfringens*) in a microwave oven under recommended standard conditions (cooking to an end-point temperature of 76.6 °C using a

2450 MHz oven) did not entirely destroy any of three kinds of bacterium. A study evaluated that internal temperature would not be a satisfactory guide for ensuring safety since the extent of bacterial growth from all microwave-cooked samples of chicken surface-contaminated by *Salmonella typhimurium* was greater than that of samples cooked in a conventional oven (Lindsay et al. 1986). Assessments were also made of the inactivation by microwave energy of pathogenic bacteria in frozen foods. Species of *Salmonella cubana*, *Staphylococcus aureus*, *Bacillus cereus*, and *Clostridium perfringes* were separately inoculated into a basic carbohydrate food, mashed potato, which, after freezing, was cooked in a microwave oven. All species survived the cooking cycle that was least detrimental to the food's palatability. In potatoes with a higher moisture content, bacteria survival was generally increased. Higher protein or lipid contents tended to allow increased survival of *Clostridium perfringes*, although higher internal temperatures were produced. Spore-forming organisms were generally less susceptible under all test conditions. Addition of whole peas to the mashed potato, to provide heterogeneity, resulted in higher temperatures, but significantly reduced microbial destruction, perhaps by allowing formation of micropockets of cooler food matrices (Spite 1984). Various cooking procedures (roasting, braising, stewing, and microwave cooking) applied to turkey thighs were evaluated for their effectiveness in removing *Campylobacter jejuni*. Roasting, braising, and stewing were effective in the destruction of *Campylobacter jejuni* in contaminated turkey thighs, even when the meat was undercooked and reached an internal temperature of only 55 °C. Destruction of *Campylobacter jejuni* by microwave cooking was assured more fully if a meat thermometer was used to check the internal temperature of the sample rather than visual evaluation (Acuff 1986). Conventional oven cookery was more effective than microwave oven cookery for reducing numbers of aerobic microorganisms and *Clostridium perfringens* in ground-beef patties when the meat was heated to approximately the same internal temperatures of 65–70 °C for rare, or 77–93 °C for well done (Wright et al. 1986). On the other hand, the extent of damage to RNA during sublethal microwave heat injury of *Staphylococcus aureus* FRI-100 was greater than that occurring as a result of conventional sublethal heating. Although both heating modes resulted in the destruction of 16S RNA, only those cells heated with microwave energy suffered lesions in their 23S RNA. When cells were allowed to recover, the conventionally heated cells regained normal profiles of 16S and 23S RNA after 180 min on the recovery medium. Microwave-heated cells restored their 16S RNA profile after 180 min of recovery, but required 270 min to restore 23S RNA with sedimentation properties similar to those found in the unheated cells (Khalil and Villota 1989).

One explanation for the better survival of bacteria in foods processed in microwave ovens is the usually short heating time and lack of microwave field uniformity. The nonhomogeneous field distribution may result in wide

Applications of Microwave Heating in the Food Industry

ranges of final internal temperatures of foods processed in microwave ovens. Use of insulated wrapping material has been suggested as one method to equalize the final internal temperature and to increase the efficiency of microbial destruction. *Clostridium perfringens* and *Salmonella typhimurium* destruction were indeed enhanced when whole turkeys were cooked in "brown-in-bags", even though the final temperature only reached 68.3°C. Survival of *Clostridium perfringens* was positively related to the number of initial spores in cooked unstuffed turkeys or in turkeys containing inoculated stuffing (Aleixo et al. 1985). Viable trichinae were found in meat paste, roast pork, and chops which were cooked in household microwave ovens (Carlin et al. 1982; Zimmermann and Beach 1982; Zimmermann 1983a). However, viable trichinae have not been found in contaminated meat microwaved in plastic roasting bags, indicating that cooling losses due to evaporation of surface moisture are the major factor in trichinae survival. Smaller selected cuts, such as boneless loin, center loin, and possibly loin roasts, should be recommended for cooking by microwave (Zimmermann 1983b). However, with a proper cooking time – power relationship, destruction of *Trichinella spiralis* can be achieved (Zimmermann 1984). The effect of a polyvinylidene chloride wrap on the range of final internal temperature and on survival of bacteria in microwave-processed foods (2450 MHz; 630 ± 50 W) was also investigated. Microbiological counts of wrapped and vented, and unwrapped foods, chicken drumsticks and ham slices, superficially inoculated with *Staphylococcus aureus* and aerobic bacteria of a few other experimental products, were made before and after microwave processing. Although wrapping did not have a statistically significant effect on final internal temperature or on counts of bacteria per gram of product tested, wrapping products generally provided a slight improvement in microbial quality when mean counts were considered (Sawyer et al. 1984). Obviously, the temperature and length of the microwave treatment are the crucial factors responsible for the lethal effects. Depending on the heating conditions, hot steam issuing from the cooking may be a more likely reason for reduction in bacterial contamination than absorption of energy itself.

In microwave heating, more than in any other heating technique, the product composition is important for the amount of microbial reduction. The competitive absorbing effect of food constituents might spare microorganisms located in the regions of minimal absorption. Thus, microorganisms which are in cold regions on the metal cavity walls or on food containers might remain unheated and viable. Furthermore, the habitual microwave heating and, in particular, the process of warming precooked food may cause a stimulation of bacterial spores and enhanced germination, leading to higher counts after heating (Blanco and Dawson 1974; Craven and Lillard 1974).

Muscle pyruvate kinase activity was established as a biochemical indicator of temperature attained during cooking in a canned cured pork product.

134   Microwave Radiation

Canned product attaining a core temperature of 62.9 °C had high pyruvate kinase activity. This activity progressively declined in a product heated to 65.6 and 68.6 °C, to become undetectable in a product processed to 69.9 °C (Davis et al. 1988). Heating patterns and temperature gradients in microwaved cooked pork could be quantified by infrared thermography (Bakanowski and Zoller 1984).

The question of whether a living system, microorganisms, for example, exposed to microwave radiation may be expected to show biological changes beyond the ones due to thermal effects of microwave energy was a matter of debate. However, no acceptable scientific hypothesis or experimental results would support athermal effects of microwave radiation. Microorganisms in freeze-dried materials, which do not heat up, survive exposure to microwave energy. Attempts to reduce bacterial populations on dried media, again under conditions which imply no heating (Grecz et al. 1964; Latimer and Matsen 1977), such as dry soil (Vela et al. 1976; Vela and Wu 1979) or spices (Vadji and Pereira 1973), failed. Such results prove that microwaves themselves are not, like ionizing radiation, specifically lethal. Almost identical death rates of *E. coli* (Brown and Morrison 1954; Carrol and Lopez 1969; Hamrick and Butler 1973), *Saccharomyces cerevisiae*, and *Streptococcus faecalis* were found for equal time–temperature relationships in water-bath and dielectric heating.

The processes of pasteurization and sterilization involve the partial or total destruction of microorganism populations in a food product, with a minimal effect on its composition. Destruction of microorganisms by microwave heating has been extensively studied practically on any kind of food which requires pasteurization or sterilization in its processing line. Essentially, food products, either packaged or unpackaged, are exposed to a microwave field in a continuous or batchwise manner (Woodburn et al. 1962; Olsen et al. 1966; Hamid et al. 1969; Kenyon et al. 1971; Hamrick and Butler 1973; Ayoub et al. 1974; Blanco and Dawson 1974; Craven and Lillard 1974; Culkin and Fung 1975; Metaxas 1976; Ockermann et al. 1976). Some of the reported attempts to pasteurize or sterilize food products are summarized in Table 18.

Information has also been published on the destruction of micrometabolites, such as aflatoxins, in peanuts (Pluyer et al. 1987).

The potential use of a microwave oven for the sterilization of baby-feeding equipment was evaluated, and guidelines for such use by health-service workers and microwave oven manufacturers were delineated. Naturally contaminated bottles and, later, nipples were exposed to moist heat in a 600 W, 2450 MHz microwave oven, followed by microbiological evaluation of the degree of bacterial destruction relative to control conditions. The results indicated that a satisfactory microwave oven sterilization method could only be reliably attained by microwave boiling for 3 min. This method, however, has few advantages over conventional sterilization

Applications of Microwave Heating in the Food Industry 135

Table 18. Partial compendium of foods pasteurized or sterilized by microwaves

Food product	References
Meat, poultry, fish and products	Watanabe and Tape (1969), Bengtsson et al. (1970), Baldwin et al. (1971), Kenyon et al. (1971), Chen et al. (1973), Ayoub et al. (1974), Blanco and Dawson (1974), Bunch et al. (1976), Crespo and Ockerman (1977), Cunningham (1980), Harrison (1980), Harrison and Carpenter (1989a,b), Schnepf and Barbeau (1989)
Milk and milk products	Hamid et al. (1969), Jaynes (1975), Tochman et al. (1985), Young and Jolly (1990)
Human milk	Sigman et al. (1989)
Goat milk	Thompson and Thompson (1990)
Soy milk	Bouno et al. (1989)
Fruit and vegetable products	Lin and Li (1971), Dahl et al. (1981)
Bread, pasta, cakes, starch	Olsen (1965)
Ready-cooked meals	Ollinger and Matthews (1988)
Eggs and related products	Rosenberg and Bohl (1987)
Cereals, seeds, spices	Vajdi and Pereira (1973)

methods and is uneconomical with respect to energy and time costs (Biela and McGill 1985).

Tempering and Thawing. Freezing extends appreciably the storage time of many foods. Naturally, frozen commodities must be warmed before use. One of the most successful industrial microwave food process is the "tempering" of frozen meat products. Tempering is the process of heating frozen foods to a temperature of $-4--2\,°C$, just below the freezing point of water. While tempering avoids problems associated with complete thawing, it does not restrict further preparations since the food is still firm and can easily be further processed by cutting, boning, etc. Microwave tempering is faster than conventional heating and can be done under more sanitary conditions, in a refrigerated room and without removing the food from the packing case. As a result, microwave tempering minimizes microbial growth and spoilage as well as weight loss, because of better juice retention. Since tempering is carried out at a temperature range in the food at which the specific heat is very low (less than 0.4) the energy required to bridge the temperature range from the storage temperature $(-24\,°C)$ to a temperature just below freezing point $(-4\,°C)$ is relatively small. Because of the very low dielectric loss of frozen food, heating is also very uniform.

Microwave tempering installations usually work at a frequency of 2450 MHz and sometimes at 896 or 915 MHz. The lower frequency might be more appropriate for thawing since it enables a deeper penetration.

Frozen foods can present unique heating problems due to different microwave absorptions and reciprocally related penetrations in ice, water, and brine. Water absorbs microwaves much more strongly than does ice and therefore, in a frozen mass, the thawed areas which absorb more energy may become overheated while still being adjacent to frozen parts. Furthermore, if the product had a history of accidental temperature fluctuation which allowed thawing and refreezing during frozen storage, the pockets of refrozen water would contain a higher salt concentration; this would absorb more energy and thus be susceptible to overheating. When microwaved, these spots of frozen water would thaw first, because of the increased salt concentration and the higher dielectric loss factor. An additional factor is the dependence of dielectric properties on temperature. Since the relative dielectric loss of a material increases with temperature, the penetration depth is also a function of temperature. As the temperature approaches $-1\,°C$, the energy needed per degree of temperature rise decreases considerably and, therefore, the risk of overheating of the food surface ("runaway") is greater. Due to all these reasons, it is virtually impossible to thaw frozen products by microwave heating without "run-away" heating. This undesired effect can, however, be attenuated if the radiation power is reduced in a duty cycle permitting thermal equilibration. Unfortunately, this procedure eliminates speed as the major advantage of microwave treatment. Circulation of refrigerated air during microwave heating could also be used to cool the product surface, and prevent surface thawing with the accompanying effects of increased energy absorption and overheating.

Because of the problems associated with complete microwave thawing, tempering is more attractive. Following microwave tempering, thawing can be done by conventional heat transfer.

Some of the products considered for thawing and tempering are mentioned in Table 19.

Drying. Drying is the removal of moisture from a product and is caused by lower water vapor pressure in the surrounding space. Historically, drying is the oldest preservation method, going back to the early periods of civilization, and it is still one of the most common means for prevention of

Table 19. Suggested uses for microwave thawing and tempering

Food products	References
Meat, meat products, poultry	Sanders (1966), Bialod et al. (1978)
Fish	Ammerman and Adres (1973)
Butter	Rosenberg and Bohl (1987)
Fruit and fruit products	Weil et al. (1970), Phan (1977)

Applications of Microwave Heating in the Food Industry 137

spoilage of food commodities. Traditionally, drying is accomplished by heating under the sun, or with any artificial source of heat, at atmospheric or reduced pressure. It is essential to maintain an optimal temperature gradient inside the material to prevent crust formation or surface burning and, at the same time, achieve a reasonable drying rate. In principle, since in microwave drying the core of the product may be heated simultaneously with the surface and there should be no need for transfer of heat from the surface to the center, the time saved seemed an attractive asset. However, the variable water content during drying makes the absorption of microwave energy in the target product nonlinear, and drying solely with dielectric heating creates a serious risk of damage by overheating. For this reason, in most applications, a combination of microwave with conventional heating, such as hot or cold air or infrared heaters, has been advised. Different heating techniques may be applied simultaneously with microwaves or in subsequent steps. The microwave drying installations are generally operated at 2450 MHz.

When drying is performed under vacuum, the temperature needed to create a drying gradient of water vapor pressure between the product and surrounding space is lower. Therefore, vacuum drying is recommended for desiccation of heat-sensitive food products, like those characterized primarily by aroma parameters. In combining microwave with vacuum drying, the arc discharge limits the intensity of the electric field which can be applied. Arc discharges were observed at water vapor pressures of less than 0.1 mm Hg for an electric field intensity of 400 V/cm. The critical discharge voltage is lower for a lower radiation frequency (915 MHz versus 2450 MHz).

The literature abounds with suggestions for drying foods by microwaves. Some representative examples are summarized in Table 20.

Table 20. Suggested uses for microwave drying of foods

Food product	References
Pasta	Maurer et al. (1972)
Other cereal and seed products	Aref et al. (1969), Fanslow and Saul (1971), Hamid et al. (1975), Roberts (1977)
Fruits and vegetables	Aref et al. (1969), Bhartia et al. (1973), Huet (1974), Rzepecka-Stuchly (1976), Huang and Yates (1980), Park (1987)
Peanut flour	Pominski and Vinnett (1989)
Milk	Souda et al. (1989)
Potato chips	Rosenberg and Bohl (1987)
Herbs and spices	Rosenberg and Bohl (1987)
Instant powdered drinks (tea, coffee)	Rosenberg and Bohl (1987)
Meat, fish, egg products	Rosenberg and Bohl (1987)

138   Microwave Radiation

The physicochemical properties and cooking quality of grains after microwave – vacuum drying treatments were not significantly different from air-dried controls. Thus, microwave techniques could be used to dry rice without materially altering the cooking and eating characteristics (Wadsworth and Koltun 1986).

Baking. The microwave baking of breads, cakes, and pastry has focused a lot of attention (Lorenz et al. 1973; Tsen et al. 1977; Tsen 1980; Martin and Tsen 1981; Hulls 1982). Microwave baking can be done at either low or high frequencies (915 and 2450 MHz), but the lower frequency might be advantageous since the deeper penetration ameliorates the risk of an unbaked core. The main disadvantages of baking with microwaves are the lack of crust formation, the inability to caramelize the surface, and, to some extent, a poorer development of aromatic components. Occasional uneven baking and tough products have also been observed. Therefore, microwave baking is often combined with conventional baking by gas or infrared heating, or with hot fat frying. The different heating mehods may be applied either simultaneously or consecutively.

Microwave baking may provide a final product with a higher nutritive value than that provided by conventional baking. The destruction of lysine, due to Maillard condensation with sugars, was attenuated in bread microwave-baked at 13.6 MHz, yielding a superior protein constituent (Tsen et al. 1977; Tsen 1980). On the other hand, microwave baking may make possible the production of bread from wheat which cannot be used for conventional bread-making, such as a flour low in protein but with a high α-amylase content. The high α-amylase content leads to extensive breakdown of starch under normal baking conditions and this in turn results in a low yield of dough which is also too permeable to gas to give an adequate volume. Furthermore, this dough lacks water-holding capacity, possesses less elasticity, and has a sticky texture. However, doughs made from low-protein wheat flours were successfully baked in a conventional oven at 320 °C simultaneously with microwave power at 896 MHz. Since in microwave baking the entire loaf is heated rapidly, the α-amylase enzyme is inactivated before its activity can progress too far and the gas development (CO_2 and steam) in the dough is accelerated. The resulting bread has a relatively high specific volume despite the low protein content. The industrial application of this procedure might, however, be delayed by the difficulty of finding a suitable nonmetal pan of acceptable food-contact material that could tolerate the high temperatures of the conventional oven.

An attractive potential is presented by rebaking, by microwaves, of conventionally baked goods. Actually, by a more accurate definition, this is a reheating treatment intended to revive frozen or stale bread products (bread loaves, pitas).

Applications of Microwave Heating in the Food Industry 139

Proofing is the step in-between panning and baking bread. The treatment calls for maintenance of the dough under controlled conditions of temperature and humidity for a final development of gas in the structure. The proofing of yeast-leavened dough fulfills several functions in the baking process: generation of leavening gases by the metabolic activity of yeast which, in turn, causes the expansion of the dough and the development of the gluten protein, formation of a shape-retaining skin, and evolution of some of the flavor components of the dough (Schiffman 1986). Microwave heating during dough proofing may cut substantially the time needed for gas development and adhesive formation.

Blanching (enzyme inactivation). Although the term 'blanching" is used liberally to describe minimal thermal treatments for a variety of purposes, it is mainly meant to describe inactivation of endogenous enzymes in living foods. Inactivation of enzymes prevents discoloration and development of unpleasant flavors in frozen fruits and vegetables. Traditionally, the blanching process is accomplished by immersing vegetables in steam or hot water. Since the essential requirement for a good blanching technique is a uniform heat distribution to the individual units of the product, microwaves present an attractive alternative.

In general, the reduction of enzymatic activities, among other biological parameters tested for effects of microwave irradiation, has focused attention (Chipley 1980). In this case no athermal effects were seen. Several studies on enzyme inactivation in fruits and vegetables have been reported (Dietrich et al. 1970; Huxsoll et al. 1970; Avisse and Varoquaux 1977). It is still debatable as to whether microwave blanching provides any advantages compared to the conventional process.

Insect Control. The different dielectric properties of cereals and various pests create the possibility for insect control in a carrier material by a selective heat shock, as an alternative to chemical insecticides. Insects can be killed by being exposed to radio frequency or microwave fields without damaging the surrounding grain (Hamid et al. 1968; Boulanger et al. 1969; Hamid and Boulanger 1969; Nelson 1976; Watters 1976; Baker and Doly 1977; Nelson and Payne 1982). Adult insects tend to be more susceptible to exposure than the immature forms. In addition to controlling the insects, the grains are also dried. The extent of grain heating is directly related to the moisture content.

The stress induced by microwave radiation to complement or potentiate the effectiveness of gamma radiation in insect destruction in grain has also been studied (Tilton and Brower 1985). Kirkpatrick et al. (1973) compared the use of gamma, infrared, microwave, and a combination of gamma with either infrared or microwave radiations in controlling *R. dominica* in soft

140   Microwave Radiation

winter wheat. The authors reported reductions of emergence of 54, 55, and 42% for gamma, infrared, and microwave treatments respectively. The combinations of gamma and infrared or gamma and microwave radiation produced significantly greater reductions of 95 and 89% respectively. Other stored-product insects may be susceptible to microwave radiation: for example, *Tribolium confusum* in wheat and flour (Watters 1976). Internal molds were reduced by 80% in freshly harvested corn kernels by microwave heating (Nolfisinger et al. 1980).

Cooking and Other Applications. In spite of earlier hopes, industrial microwave cooking has, as yet, found only limited application. One of the most noteworthy uses is in precooking bacon. Microwave preparation of bacon controls the product shrink, allows recovery of rendered fat, and produces a uniform, high-quality precooked product with virtually no waste and cooking odors. The effect of microwaves on the nitrite salts, which are used as preservatives, has been evaluated since the cooking of bacon containing nitrites could potentially lead to the formation of nitrosamines (Miller et al. 1989). Bacon was analyzed for volatile nitrosamines after microwave cooking, and the results were compared with those obtained after frying bacon in a pan. Microwave cooking gave statistically significantly lower levels of all three volatile nitrosamines detected in the bacon (Osterdahl and Alriksson 1990).

Microwave cooking has been established much better in the home than in industry because of its speed and convenience. This way of cooking is composed of two heating phases: instantaneous heating due to radiation and a delayed heating by conduction inside the food mass. Measurements of temperature rise in water, chicken frankfurters, and cake cones after conventional hot air processing and after microwave processing indicated that temperature rise occurred more often in products heat-processed in microwave ovens than in hot air ovens (Sawyer 1985). However, the shorter duration of microwave cooking reduces the time–temperature exposure of microoraganisms in food. Microwave cooking of meat loaf, for example, resulted in a lower reduction of the microbial population compared to cooking in conventional ovens or in crock pots (Fruin and Guthertz 1982).

Lipid oxidation is one of the major factors in quality deterioration of meat products and NaCl, which is added to several poultry products for its functional properties, might promote this type of degradation. It appears that heating by microwaves may retard the prooxidizing effect of NaCl in convenience items, such as ready-to-serve turkey meat patties (King and Bosch 1990).

Testing panelists rated chicken (Barbeau and Schnepf 1989), restructured beef steaks (Penfield et al. 1989), and lamb roast (Ray et al. 1985) cooked in the conventional oven as more tendar and juicy than that cooked in microwave ovens. Microwave cooking produced lower evaporative losses

Applications of Microwave Heating in the Food Industry 141

and higher drip losses in pork tenderloin steaks (Prusa and Hughes 1986). Microwave-heated flounder fillets were significantly more crumbly than those heated conventionally. However, the panels did not detect differences due to heat treatment in the flavor, appearance, or overall acceptability of either species (Brady et al. 1985). Greenland turbot fillet sections were cooked faster in a microwave oven than in a conventional oven. Solid drip and total cooking losses were greater and shear values and evaporative loss were lower for microwave-heated fish. Turbot cooked in the microwave oven was softer and less chewy than was conventionally heated fish, but was judged to be equally flaky and moist (Madeira and Penfield 1985).

Food and nutrition scientists have been trying to use a variety of methods to decrease the amount of fat in the diet. Fat, the most calorie-dense food constituent, is a major factor in weights gain. A high intake of saturated fat and cholesterol has been shown to be a risk factor for various cardiovascular diseases and cancer. Ground-beef patties were cooked by different methods (broiling, charbroiling, roasting, convection heating, frying, and microwave) and analyzed for compositional changes. For cooking methods other than broiling or convection heating, yield decreased with increasing fat levels. When broiling, charbroiling, roasting, or frying were used, patties with 14% fat incrased in fat retention, while patties with 24% fat decreased in fat retention. Microwave cooking reduced fat content in all patties (Berry and Leddy 1984). Steaks, however, generally had more protein and less fat and calories when cooked by broiling compared to cooking by microwaves (Berg et al. 1985). Among the cooking procedures to decrease fat content the use of fat-absorbing materials during cooking has received attention. This approach is particularly suitable with microwave ovens because of selective heating action. Currently, paper towels are generally used to absorb grease from food products. Ultrafine polymer fibers, such those of polypropylene, melt-blown into a white web, have been used to clean marine oil spills. The same principle of fat absorption may be applied when heating foods in microwave ovens, with the additional advantage that this material, being hydrophobic, would repel moisture loss. Indeed, bacon and precooked sausage patties were microwave heated on melt-blown polypropylene pads or on paper towels. Panelists scored the food heated on the polymer pad as less greasy and chewy, and less fat and more moisture were present in the high-fat meats cooked on the melt-blown pad (Costello et al. 1990).

The baking of potatoes exemplifies the potential effects of different rates for enzyme deactivation during cooking. Sweet potatoes baked in a microwave oven have a drier feel in the mouth than those baked in a convection oven. A comparison of the carbohydrate fractions showed that the samples baked in the microwave oven contained less total alcohol-soluble carbohydrates, reducing sugars, and dextrins and more starch than those cooked in the convection oven. Sweet potatoes contain amylolytic enzymes, such as

α-amylase, that hydrolyze starch when roots are cooked. Differences in the amount of α-amylase activity appear to account for differences in the feeling of moistness or dryness in the mouth. Apparently, microwave cooking was too rapid to allow significant starch degradation by α-amylase. It is, however, possible that a schedule of cooking could be devised such that the sweet potato would be heated to 80–85 °C, maintained at that temperature for about 7 min, and then cooking finished at full power. Such a regime could result in a microwave-baked sweet potato of quality similar to that prepared by convection baking (Purcell and Walter 1988).

Foods cooked by microwaves are as nutritious and, in some instances more nutritious, than those cooked conventionally. Thus, thiamin retention on a dry weight basis ranged from 77% in conventionally cooked chicken breasts to 98% in microwave-cooked chicken legs (Barbeau and Schnepf 1989). Low-power techniques for microwave ovens yield equal or better retention of thiamin, riboflavin, pyridoxine, folic acid, and ascorbic acid in comparison with conventional methods. Sensory qualities of microwave-cooked foods compare favorably with those of conventionally cooked food (Hoffman and Zabik 1985). However, fish heated in a conventional or microwave oven to an end point of 70 °C did not differ in thiamin content (Brady et al. 1985). Composition of broiler meat is influenced by cooking methods. Crude protein was higher in meat cooked by microwaving, deep fat drying, and pan frying than in meat that was baked or broiled. Crude fat and the minerals Ca, P, K, S, Cu, and Mn were not affected by the cooking method. Broiling was the most destructive cooking method to Fe and Zn retention and in microwave heating more Mg, Fe, Zn, and Na were retained. Higher mineral concentrations were found in uncoated meat than in coated meat (Proctor and Cunningham 1983).

Recently, interest has been growing in fish and fish products as sources of polyunsaturated fatty acids, mainly of the ω-3 family. This interest stems largely from studies which suggest that ω-3 polyunsaturated fatty acids may play an important role in prevention and management of cardiovascular diseases. This prompted an evaluation of the stability of oils high in polyunsaturated fatty acids in fresh fish of low, medium, and high fat content cooked using a microwave oven. The effect of cooking was minimal with no detectable difference in total lipid content between cooked and uncooked samples. Most important, polyunsaturated fatty acids were virtually unaffected by the overall cooking and the cooked fish retained the original composition and content of polyunsaturated fatty acids (Hearn et al. 1987).

Soybeans are one of the best sources of high quality plant protein and edible oil. However, they also contain various chemical factors which elicit adverse biological responses in animals. These effects are interrelated: growth inhibition, reduced fat absorption and protein digestibility, and pancreatic hypertrophy. Traditional methods of heating, such as toasting, roasting, and heating in boiling water, are effective ways to improve the

Applications of Microwave Heating in the Food Industry 143

nutritional value of soybeans. Therefore, the effects of microwave heating on soybeans have been evaluated. Microwave heating decreased the solubility of soy protein. It also decreased significantly the phytate and phospholipid contents of soybean seeds and, as a result, the content of inorganic phosphorus increased (Hafez et al. 1985, 1989). Furthermore, microwave treatment of dry soybeans caused a significant loss of triglycerides containing more than four double bonds. This loss could be diminished if water-soaked soybeans were microwaved. The heating period needed for inactivation of trypsin inhibitor was also shorter for water-soaked beans (24.3% moisture) compared to unsoaked soybeans (Yoshida and Kajimoto 1988). Soybeans and their products are also relatively good sources of vitamin E (tocopherols) which is considered to be an important nutritive factor, though the mechanism of action is not yet fully undestood. The amounts of α-, β-, γ-, and δ-tocopherols in soybeans before microwave treatment were in the range 6.2–13.0, 2.7–4.5, 60.0–76.8, and 45.7–57.9 mg per 100 g lipid respectively. During microwaving oxidative degradation of tocopherols may occur. However, a microwave treatment of soybeans which would be optimal for preparation of full-fat soyflour without a burnt odor retained approximately 90% of the inidividual tocopherols with a few exceptions and caused no significant difference in the chemical changes of the lipids (Yoshida and Kajimoto 1989). Surprisingly, the reduction of tocopherols in microwave-heated soybean, compared to linseed, corn, olive, and palm oils, was not directly related to the chemical properties of the oils (Yoshida et al. 1990).

Technological problems in the processing of other dry legumes are similar to those for soybeans. Thus, dry beans also constitute an important source of inexpensive food protein but their consumption is limited due to their association with flatulence and inconvenience created by the "hard-to-cook" phenomenon, which includes long soaking and cooking times. Efforts to improve the cooking quality of dry beans have included various pretreatments, including heating with microwaves (Esaka et al. 1987) and also with infrared radiation (Kadir et al. 1990). Lipoxygenase and trypsin inhibitors were completely inactivated in dry, and even faster in soaked, whole winged bean seeds.

Among microwave food products, microwave popcorn has been successful in the retail market in spite of problems such as low expansion volume, a large number of unpopped kernels, and scorching of the popped kernels. The microwave behavior of popcorn depends on several factors. Each variety of popcorn has its optimum moisture content for the ratio of maximum expansion volume to mimimum unpopped kernels. A long-term frozen storage increases the popped volume compared with storage at room temperature. Not surprisingly, the integrity of the pericarp is very critical in popping performance and, therefore, surface damage to kernels reduces drastically the expansion volume. Likewise, salt and oil affect the popping performance of popcorn in a complex manner (Lin and Anantheswaran 1988). Kernel size also correlates with expansion volumes during popping. A greater popping volume was observed in varieties having sphericity values greater than 0.70. For a given variety, increasing kernel size resulted in a reduced ratio of unpopped kernels to expansion volume. The elastic deformation had no correlation with microwave popping (Pordesimo et al. 1990).

The formulation of composite dishes could be adapted for microwave heating. Some conventional recipes are easy to tailor. These are generally high-moisture-type foods or foods that are steamed, poached, covered, or cooked in sauces. Even these formulations require changes in cooking time and cookware used. Other recipes are more difficult to adapt to microwave heating; these include egg and cheese dishes, which tend to get stringy or tough in texture, and sauces, which may curdle. Greater tolerance and structural control may need to be built into baked goods. These foods may require changes in ingredients and size.

Food cooked in microwave ovens does not brown, as its surface does not reach temperatures required for Maillard reactions and, therefore, foods that normally brown need browning agents or sauces to disguise them. An alternative approach to solving this problem has been the use of susceptor trays which work on the principle that localized heat from microwave-coupled metal is transferred to food surfaces in immediate contact.

Finally, there are foods which, by the nature of the ingredients, final appearance, or consistency, are not suitable for microwave cooking. Included here are: deep-fat fried foods which may reach extremely high temperatures due to the absorption of microwaves by fats; foods requiring a crisp crust or dry surface which may become soggy; and foods requiring dry hot air for leavening.

In conclusion, it seems that cooking with microwaves confers no qualities to food other than those due to efficient heating. Lethal effects on microorganisms and trichinae are due to heat. Speed and evenness of heating are influenced by the composition and mass of food, as well as by features of the appliance. Since the heating during microwave cooking could be uneven, the presence of relatively cool regions might account for the survival of bacteria when very high temperatures are recorded in other parts of a food (Dealler and Lacey 1990). The use of standing time after cooking is important to allow the even distribution of heat by conduction, without overcooking. The formulations of microwavable dishes should be based on the understanding of microwave heating mechanisms. The uniformity of size and shape, which ensures an even penetration of the microwaves is critical. The microwave oven is primarily a steamer and works well for foods that can be steamed, but it is almost impossible to crisp a food in a microwave oven without an auxiliary heat source. Problems unique to microwave cooking include requirements for rotating, stirring, and mixing food during preparation to minimize hot and cold spots, the proper use of different power settings, the loss of moisture as a function of power output, the requirement for round, rather than rectangular, food containers, and the increase in heating rate with increased salt content (producing surface overheating). Athough there is variability, microwave ovens have considerable potential for energy saving.

Premature enthusiasm, or even euphoria, regarding microwave treatments should not cloud the facts, already evident to discerning scientists, that technical difficulties still exist, along with potential problems of unknown dimensions. Some present techniques or operational designs may be inadequate to deal with microwave treatments. Limitations should be recognized, otherwise elevated expectations will be deflated by realities.

4.5 Materials for Food Containers for Microwave Treatment

The heating of convenience foods with microwaves has posed new questions on the compatibility between packaging materials and food. Intensive research and development efforts have been invested to adapt materials for the unique features of microwave heating since the container becomes an active component in a microwavable product. Fractions of the microwave energy dispensed in the oven are absorbed by, reflected from, or transmitted through the product and the container.

By its nature, microwave heating precludes the use of metals. Any packaging material which is made of metal or which contains any metallic components will cause severe arcing and constitutes a serious hazard. A priori, glass would be the best material for food containers exposed to microwaves. It is transparent to microwaves and can with stand high temperatures, either of the food or of the combination of conventional and microwave ovens. Unfortunately, its use is restricted by its fragility and heavy weight.

Within the wide range of synthetic polymers available some are more suited to microwave exposure than others. Polystyrene (styrofoam) is an excellent insulator but distorts at 77 °C. It is used to microwave-heat sandwiches and beverages just before serving. Polyethylene is quite brittle and becomes more so at freezer temperatures; it distorts and melts at 77 °C and absorbs some microwave energy. Polypropylene can resist temperatures up to 100 °C but does not freeze well. Melamine, although it has an even higher distortion temperature, is not suited for microwave heating because of its significant energy absorption. Polysulfones and thermoset-filled polyesters are indifferent to temperatures in the range $-18-200$ °C, have only a minimal absorption of microwaves and, in spite of a more expensive price, are extensively used in microwave cooking. Coated paper and cardboard containers are suitable for warming up precooked foods (Annis 1980).

Convenience foods that were packaged for conventional preparation are now appearing in containers suitable for use in microwave or conventional ovens, or are even designed solely for use in microwave ovens. A few specific challenges are associated with the development of microwavable foods, such as preservation of crispness and browning. Typically, when a product that is meant to be crispy is heated in a microwave oven it releases moisture which, unless dissipated from the product environment, softens the coating. Therefore, modifications in package design were made to facilitate the removal of moisture. The packaging of convenience dishes has also became an integral part of the cooking process. A heater board, in which a heat susceptor is laminated onto paperboard, helps to produce browning and crisping. Such a board controls the amount of microwave energy that impinges upon the board and upon the product. The microwave energy is locally converted by the heat susceptor into heat which is applied to the product surface it contacts. Thus, the dielectric heating of the food inside such packaging is complemented by the board which actively generates heat to brown and crisp the food. The heat susceptors are small pieces of metal, usually aluminum, that are vacuum deposited or splatter-coated on a plastic film, and then laminated into the paperboard.

The novel materials used in microwave cooking present unfamiliar questions about their interaction with food. Therefore, packaging materials intended for use in microwave ovens must be tested to determine their suitability for this method of heating, the extent of shielding they provide, and whether the heat causes degradation and eventual contamination of food. The extreme conditions of misuse should be taken into account (Booker and Friese 1989). It is beyond any doubt that products of the interaction between microwaves, food and packaging are due only to the heat generated and not to any other effects of radiation.

When a material is heated two classes of volatiles are released: thermally desorbed compounds, which are indigenous to the material, and products produced from the pyrolysis of the packaging material. Compounds which might be thermally desorbed include residual chemicals from the manufacturing process and solvents from adhesives used. Products of pyrolysis may derive from the packaging material itself, such as coatings or inks. Modern analytical assays can detect the release of these volatile compounds in trace amounts of nanograms per square inch. The origin of the released volatiles is decided from plots of volatile yield vs temperature. This dependency is distinguishably different for absorbed contaminants and for decomposition products. The adsorbed volatiles begin to desorb rapidly and the amount of desorbed compound quickly reaches a maximum concentration. By comparison, pyrolysis volatiles begin to appear at elevated temperatures and are continuously generated without reaching a limit.

In paperboard products, compounds such as 1,1,1-trichloroethane, toluene, low molecular weight aliphatic hydrocarbons, xylene, naphthalene,

Safety Aspects of Microwave Heating Equipment 147

methylene chloride, fluorocarbons, and styrene may be found as artifacts of the papermaking process. Furfural, which is generated by the decomposition of pentosans, is one of the primary pyrolysis products from paperboard. Although it is neither toxic nor particularly obnoxious, it imparts a distinctive, extraneous taste to the food products it contacts. Several factors, particularly the pentosan content of the paperboard, acidity, temperature, and time of exposure, influence the amount of furfural found among the volatiles in microwaved paperboard products. Alcoxyalcohols such as 2-butoxy-1-ethanol, low molecular weight alcohols, acetic acid, toluene, and 1,1,1-trichloroethane are but a few of the compounds originating from adhesives. Benzene arises from the pyrolysis of polyvinyl acetate used in the adhesive. Printing ink is the source of styrene and α-methylstyrene.

Migration of chemicals from plasticized films into foods increases with direct contact, with increasing temperature, and with contact time (Castle et al. 1988a). Epoxidized soybean oil is used as a plasticizer and heat stabilizer in a number of food contact materials, in particular in polyvinyl chloride films. When such a film was used in direct contact, to cover food for cooking in a microwave oven, levels of epoxidized soybean oil from 5 to 85 mg/kg were observed, whereas when the film was employed only as a splash cover for reheating foods the levels ranged from 0.1 to 16 mg/kg (Castle et al. 1990). Likewise, the migration of total levels of polyethylene terephthalate oligomers from roasting bags, trays (for conventional and microwave oven use), and "susceptor pads" for microwave browning applications into a diverse range of foods has been determined. Total levels of migration of oligomers were found to range from 0.02 to 2.73 mg/kg, depending on the foodstuff and on the temperature attained during cooking. On repeated use of trays there was a decline in migration of oligomers from the first to second and subsequent uses of the container (Castle et al. 1989). To reduce the migration of plasticizers from plastic films into foods two approaches were proposed; firstly, the production of thinner films with a consequent reduction in the level of plasticizer normally present and, secondly, its partial or total replacement with a polymeric plasticizer (Castle et al. 1988b).

4.6 Safety Aspects of Microwave Heating Equipment

Subconsciously, on the one hand, microwave ovens are perceived as being safer than conventional ovens because the interior does not heat up; on the other hand, the unknown hazards of exposure to an excessive level of radiation are intimidating.

Microwaves, which correspond to energy level changes of about 10^{-2} eV, are obviously insufficient to ionize biological tissues, which possess ionization potentials of several electron volts (usually more than 10). Likewise, the quantum energy level of radiation in the microwave range is many orders of magnitude lower than is necessary to break chemical bonds. Consequently, microwave radiation is not mutagenic or carcinogenic and its main effect is of a thermal nature. Within certain limits the body may absorb radiation and can adapt automatically to the resulting temperature increase by removing excess heat in the bloodflow. However, should the radiation become too intense the thermal balance could no longer be restored by bodily processes and burns would then occur. As microwaves tend to heat deep within a body, one might fear that deep burns would occur while the surface temperature remained acceptable. Naturally, avascular structures with a relatively poor thermal exchange are the most heat-sensitive organs. Male testicles are sensitive to heat and microwave radiation may produce temporary sterility. For the same reason of avascularity, thermal effects of microwaves have been implicated in development of lenticular opacities. There exists a certain radiation threshold beyond which such irreversible changes occur. The eye was found to possess a damage threshold of around $150\,mW/cm^2$ for development of cataracts after 90 min continuous exposure. A considerable number of other studies have been carried out to determine the damage threshold and no permanent effect has been observed for power levels lower than $100\,mW/cm^2$. For comparison, it is noted that the solar radiation at noon on a sunny summer day reaches a level of $100\,mW/cm^2$ in the infrared range. Introducing a safety factor of 10, the value of $10\,mW/cm^2$ was adopted in the United States and in western European countries as the upper limit tolerable for microwave irradiation of an indefinite length of time. (It should also be noted that the limit of $10\,mW/cm^2$ is very much below the power levels used in medical treatments. To obtain any kind of therapeutical effect with microwaves, the lower threshold of $100\,mW/cm^2$ must be exceeded.)

Some researchers feel that hazards may occur even at very low levels of microwave radiation. In the Soviet Union and in eastern Europe the upper limit for long-term exposure to microwaves was set at $10\,\mu W/cm^2$, which is one thousand times smaller than the limit prescribed in the United States and elsewhere. This very low limit was justified by the claim for the existence of nonthermal effects produced by other interactions of microwave radiation with the organism. Health problems attributed to nonthermal effects include nervousness, hormonal imbalance, malformations, and anomalous brain activity. However, the experimental support for connecting such symptoms to microwave radiation is questionable. The effects noticed and published by some researchers could not be duplicated by others under supposedly identical conditions. It is possible that some symptoms attributed to microwaves were actually produced by some other mechanism, for instance, X-rays emitted by a poorly shielded magnetron or ozone emanating from a defective electric supply.

A particular note of caution is related to people who have pacemakers. The presence of metallic conductors within the body can trigger other

Analytical Applications of Microwave Radiation 149

complex interactions. Microwave radiation, penetrating deep within the body, may induce a current in cardiac pacemakers, causing a hazard to their owners. Experiments were carried out with signals at several frequencies and operating regimes. The most significant effects were noted for radar signals at around 9 GHz. At the industrial 2450 MHz frequency, only small changes of rhythm have been observed up to the maximum level of the experiment, which was set at 25 mW/cm^2.

The restriction that 10 mW/cm^2 radiation should never be exceeded in the vicinity of microwave ovens has led to the need to shield very carefully all equipment, particularly the applicators, to avoid any hazard to their operators. Ovens are equipped with safety latches, which cut off the power as soon as the door is opened. Doors are further equipped with chokes and absorbing strips which reduce very significantly any leakage toward the outside. Industrial equipment is provided with energy-trapping devices at the conveyor or product openings. The great majority of ovens presently built for domestic use are carefully shielded, with their leakage, at the time of purchase, being much lower than the 1 mW/cm^2 limit at 5 cm from the door, prescribed in several countries. Although microwave ovens have been around for more than 30 years and millions of them are routinely operated, no major accident has been reported to-date, except maybe some minor burns when touching hot dishes (Gardiol 1984).

Another facet of the safety of microwave use in the food industry is the question of whether or not specific toxic components may be produced in food as a result of microwave exposure. Since microwave radiation cannot break chemical bonds, the possibility of induced chemical reactions, other than thermal ones, is nonexistent. Consequently, no compounds, toxic or otherwise, other than those generated by any conventional heating method, could be expected. Compared to the more conventional methods of cooking, such as broiling, roasting, or frying, microwave cooking is newer. As a result, the characteristics of microwave heating are not as well understood and accepted by the public as those inherent in conventional cooking. This sometimes results in controversy regarding the safety of microwave-cooked foods. Variation in food configuration (shape, size, etc.) or composition can result in underdone products. This, however, applies to any method of cooking and underscores the need for good judgment and common sense in dealing with possibly contaminated food.

4.7 Analytical Applications of Microwave Radiation

Determination of Moisture Content. Drying by microwave heating has been employed in methods for the rapid determination of moisture in a variety of foodstuffs (Pettinati 1975; Pieper et al. 1977; Shanley and Jameson 1981; Chin et al. 1985; Wang 1987). In principle, microwave radiation, which is absorbed by the water molecules distributed throughout the sample, eliminates the problems, such as overheating, crust formation, and charring, associated with other forms of rapid conventional heating. Commercial equipment for dry matter analysis consists of an electronic balance housed within a microwave oven and provided with a microprocessor giving automatic control and direct readout of the moisture content. The inexpensive domestic microwave oven can be modified for analytical use (Shanley and Jameson 1981; Andrews and Atkinson 1984). The major modification is aimed at protecting the magnetron against energy buildup when the samples become dry, by installing a small water balast as a dummy load. The containers used to hold the sample vary from fiberglass pads or filter paper to plastic dishes. As in conventional heating, the sample should also be spread out for even exposure in the microwave dryer. High sugar and protein products are better dried at lower temperatures under vacuum to avoid decomposition due to a rise in temperature. When sample particles are of different sizes, grinding insures even size and homogeneous drying. Sometimes, the addition of a polar liquid to a sample, like water or alcohol, enhances the drying rate. Finally, the incorporation of a rotating table in the oven eliminates possible burning due to hot spots, and enables testing of multiple samples.

Microwave Sample Preparation for Spectroscopic Analysis. Microwave heating simplifies preparation of samples for atomic absorption and emission spectroscopy and for other analytical procedures. The technique of microwave-heating samples in sealed containers to speed up digestion has been in widespred use for the past few years (Sturgeon and Matusiewicz 1989). Ovens specifically engineered for this purpose regulate and record pressure conditions inside digestion vessels, allowing organic and inorganic samples to be put in solution.

Analytical NMR Methods for Quality Control in the Food Industry. Nuclear magnetic resonance (NMR) spectroscopy is a technique based on the principles of nuclei polarization and serves to detect quantitatively certain chemical elements by monitoring the amount of radiofrequency energy absorbed. The main difference between NMR and other spectroscopic techniques lays in the fact that in optical spectroscopy the energy levels are intrinsic properties of the molecules resulting from their electronic, vibrational and rotational motions, while in NMR, the levels studied depend upon the presence of a strong, external, magnetic field.

Most organic and inorganic compounds exibit a nuclear magnetism which arises from the magnetic moments of the nuclei in the compound (Fig. 24).

In the natural state nuclei are oriented randomly. Application of a strong external magnetic field leads to the polarization of the spins, that is, the magnetic moments of the nuclei can align with or against the field, thus creating two separate energy levels. When equilibrium between the popu-

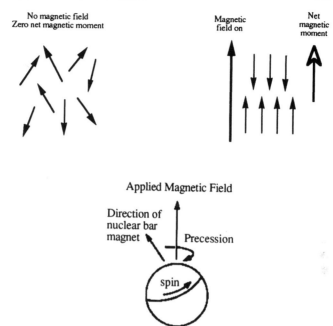

Fig. 24. Orientation of a spinning nucleus in the direction of the external magnetic field. The nucleus has magnetic moment, μ, and angular velocity of precession, ω

lations of the two states is reached, the slightly more stable alignment state with the field is more populated. This preference creates a net macroscopic nuclear magnetic moment in the sample. Once the population difference is established, it is possible to equalize the populations (that is, to rotate away the excess population from the field direction) by irradiation of the sample with energy of the correct frequency, called the frequency of resonance. The corresponding energy ranges from 10^{-2} to 10^{-6} eV, which is in the radio wave range, and indicates that this method can be used for measuring extremely low energy separations or energy splittings.

All nuclei have a characteristic magnetogyric ratio (γ) or frequency-field relationship, such that naming either the frequency or field strength in conjunction with a particular nucleus, such as hydrogen, identifies the characteristics of the magnetic field employed. Mathematically, this relationship is:

$$\omega_0 = \gamma H_0$$

or

$$\nu_0 = \frac{\gamma}{2\pi} H_0$$

where ω_0 is the precession resonance frequency in radians per second, v_0 is the resonance frequency in hertz, and H_0 is the strength of the applied field in tesla.

The proton (^1H) is probably the most common element used for routine applications of NMR. At present, magnets employed in NMR research span the range from 20 MHz (0.47 T) to 600 MHz (14.1 T). For most high-resolution structural studies, the highest attainable field is the most useful.

The nuclear magnetic resonance in bulk materials can be observed in several ways. The "continuous wave" experiment consists of slowly sweeping the radio frequencies applied to a sample in a fixed magnetic field or, alternatively, slowly sweeping the magnetic field with a fixed frequency. A spectrum, composed of small shifts in the absorption peaks, is observed and can be interpreted in terms of the various chemical environments of the particular element being studied. Splitting of some of the peaks into multiples occurs due to interactions between the various magnetic nuclei within the molecules, and this gives further information regarding the structural relationship of the atoms to one another.

Alternatively, a resonance nuclear state can be achieved with short bursts, or pulses, of electromagnetic energy of exactly the same frequency as their precession. At that frequency the atomic nuclei absorb energy and, as a consequence, the magnetic moments change their alignment relative to the applied magnetic field. It is possible to turn them through 90°, if the magnetic field and the appropriate energy are applied simultaneously as a 90° pulse. The observation of the nuclear spin system is made after the pulse is turned off. After a 90° pulse, the nuclei have surplus energy which they can radiate at the same resonant frequency to the surroundings by a process known as spin relaxation. This process can be detected and measured. The length of time needed for this reemission is associated with the environment of the nuclei because the easier it is for them to pass energy to neighboring atoms, the quicker they can return to the original state. The strength of the signal falls exponentially with time at a rate characteristic of the environment, a measurement known as the relaxation time. The time constant (T_1) is called the spin-lattice or thermal relaxation time. This relaxation involves transfer of energy from the nucleus in its high energy state to the molecular lattice. Values between 10^{-3} s to several minutes are usual. In addition, the average lifetime of the excited state of a nucleus is also controlled by spin-spin (or transversal) relaxations. In spin-spin relaxations a nucleus in an upper energy level transfers its energy to a neighboring nucleus of the same isotope by a mutual exchange of spin. This process does not reduce the total population of the upper spin state, although it does affect the lifetime of a particular excited nucleus. The efficiency of this exponential relaxation process is expressed in terms of a characteristic time, T_2. The parameters T_1 and T_2 can be rapidly and precisely determined with a pulse spectrometer, and knowledge of these values reveals a wealth of information about the sample.

Analytical Applications of Microwave Radiation 153

It is the measurement of spin-spin relaxation time following a radio-frequency pulse which has found applications in analytical food chemistry. These applications include polymorphism studies of fats, determination of oil content in seeds, grains, and dairy products, and moisture content in bakery products, cocoa products, and margarine.

Considerable attention has been devoted to the determination of the fraction of solid fat in fats (van Putte et al. 1975; Templeman 1977). Fats and oils are composed of a mixture of triglycerides of different fatty acids. The triglycerides can crystallize into different crystal modifications (polymorphism) which have quite different melting characteristics. Although this important parameter of the solid fat content can be determined by dilatometry, differential thermal analysis, or differential scanning calorimetry, it can be measured most rapidly by NMR spectroscopy. The pulse NMR method exploits the fact that the protons present in the liquid phase, like molten fat, have a higher degree of mobility than those in the solid phase. Thus, in solid fat, protons exchange their energy rapidly because of the tight contact imposed by the rigid lattice, which shortens the spin-spin relaxation time to about about $10\,\mu s$ compared to $100\,ms$ for liquid oil. This means that the proton resonance free-induction decay signal following a strong radiofrequency pulse contains two components having distinctly different decay time constants, which represent the solid and liquid components within the sample. To obtain an optimum signal-to-noise ratio, a $90°$ radiofrequency pulse is used and, by suitable sampling of the decaying signal at two different points in time, one obtains a direct digital readout of the percent of either the solid or liquid present.

The determination of oil in oilseeds by NMR (Tiwari and Burk 1980) is based upon the fact that the proton transverse relaxation time $(T_2 = 100\,ms)$ is much larger in oils than in the other main constituents of oilseeds: protein, carbohydrate, or water. This makes it possible to measure the oil in the free-induction decay signal without any interference from other parts of the seed. The signal is converted into percent of oil in seeds by using a linear calibration curve.

It is also possible to determine the water content in margarine by the pulse NMR method. However, since the relaxation times of water and oil do not differ greatly $(T_1H_2O = 1\,s; T_1\,oil = 200\,ms)$ it is necessary to dope the water with paramagnetic ions so that its relaxation time falls to $1\,ms$ to provide a sufficiently large difference to that of oil.

NMR spectroscopy is also employed to determine water content in low-fat or fat-free foods such as flour, sugar, and sweets. It is also easy to determine the bound water with NMR spectroscopy by temperature studies near the freezing point, which separates out the water which does not freeze (Leung et al. 1976).

Applications of NMR in agriculture, including NMR imaging, have been recently reviewed (Pfeffer and Gerasimowicz 1989).

Electron Spin Resonance (ESR). This form of spectroscopy (also sometimes called by the synonymous term electron paramagnetic resonance, EPR) is based on the absorption of electromagnetic radiation, usually in the microwave region, which causes transitions between energy levels produced by the action of a magnetic field on an unpaired electron. An electron may be pictured as a spinning, negatively charged particle. By virtue of its charge and spin, the electron behaves as a tiny magnet and can interact with the external magnetic field. The phenomenon is basically the same as that in NMR, except that the magnetic moment of an electron is about 658 times that of a proton, so that in a magnetic field the energy level separation for electrons is 658 times that of the protons. This accounts for the difference in the spectral regions of the techniques at equivalent magnetic fields (20–600 MHz for NMR vs 2000–35 000 MHz for ESR) and for the much greater sensitivity of ESR. Thus, while many of the basic concepts are common to the two techniques, the instrumentation and range of applications are very different.

As a ground rule, ESR spectroscopy has been used exclusively by the research scientist as an analytical tool for detection and characterization of free radicals, although the potential for application in quality control exists. Specifically, ESR offers some hope of a reliable method to detect the irradiation of food. When food is irradiated with ionizing radiation, free radicals are generated. Most of these free radicals decay instantly into stable radiolytic products, except for those trapped in hard tissues where they can remain stable for a long time. It is these surviving radicals that ESR is capable of detecting in irradiated grains, spices, or foods with shells or bones.

Chapter 5 Case Studies

5.1 Suppression of Postharvest Pathogens of Fresh Fruits and Vegetables by Ionizing Radiation

R. Barkai-Golan[1]

5.1.1 Introduction

Following the harvest, fresh fruits and vegetables become susceptible to invasion by postharvest microorganisms which generally would not attack them while they are still attached to the parent plant. The storing of fruits or vegetables for considerable periods of time to extend the marketing season and to enable their shipment to destinations distant from areas of production is accompanied by continuous physiological and chemical changes associated with ripening and senescence. Along with these changes, the vulnerability to postharvest infection increases. As the resistance to diseases usually decreases with postharvest time (Eckert 1978), economic losses attributed to microbial deterioration become heavier with longer periods of storage and transportation. The spoilage rates may be especially high in hot and humid climates, conditions which are particularly favorable for pathogen growth.

Since the 1950s, intensive studies have been conducted to evaluate the possibilities of utilizing ionizing radiation for extending the useful shelf-life of fresh fruits and vegetables. Radiation studies were aimed mainly at three directions: (1) control of postharvest diseases; (2) delay in ripening and senescence processes; and (3) control of insect infestation for quarantine purposes.

The present review will focus on irradiation treatments as a means for extending the postharvest life of fruits and vegetables by suppressing disease development. Studies in this area have been discussed in a number of reviews (Salunkhe 1961; Mercier and MacQueen 1965; Kaindl 1966; Sommer and Fortlage 1966; Sommer and Maxie 1966; Ahmed et al. 1968;

---

[1] Department of Fruits and Vegetable Storage, Agricultural Research Organizations, The Volcani Center, P.O. Box 6, Bet Dagan 50250, Israel

Dennison and Ahmed 1975; Brodrick and Thomas 1978; Moy 1983; Barkai-Golan 1985; Thomas 1985, 1986a,b). An important advantage of gamma rays is their ability to penetrate deeply into the host tissues. Thus, in contrast to chemicals, gamma radiation enables not only the control of surface- or wound-infesting microorganisms but also pathogens implanted into the host as latent or active infections. Therefore, gamma radiation may also be considered as a therapeutic means for postharvest diseases.

With the increased tendency to reduce the use of chemical applications on fruits and vegetables, the nonresidual feature of ionizing radiation is an important advantage. On the other hand, the effective use of ionizing radiation as a fungicidal or fungistatic treatment is limited by the susceptibility of the host tissues, as expressed by radiation-induced peel damage and adverse changes in color, texture, flavor, or aroma. Thus, the use of irradiation as a means for decay control will depend on the balance between pathogen sensitivity and host resistance to the treatment.

This section discusses the factors affecting pathogen susceptibility to ionizing radiation in various host–parasite systems, the relationship between ripening retardation and disease development, and methods for reducing the effective radiation dose below the threshold of damage to the host.

5.1.2 Radiation Effects on Pathogens

Ionizing radiation may damage directly the genetic material of the living cell, leading to mutagenesis and eventually to cell death. It is generally agreed that nuclear DNA, which is recognized for its central role in the cell, is the most important target molecule for radiation of microorganisms (Gordy et al. 1965; Haynes 1966; Coggle 1973). Among the many types of DNA modifications which contribute to cell death, most of the information is concerned with single- or double-strand breaks in the DNA, their yield, repair mechanisms, and biological significance (Johansen 1975).

Eukaryotic cells, such as molds and yeasts, have a relatively large nucleus surrounded by a membrane and organized into distinct chromosomes. These nuclei represent larger targets than the genomes of prokaryotic cells of the vegetative bacteria and spores. The latter are relatively small, without a specialized nuclear membrane, and with the DNA molecule apparently freely suspended in the cytoplasm. It is not surprising, therefore, that eukaryotes are generally more sensitive to radiation than prokaryotes (Grecz et al. 1983). Exceptions are the coenocytic fungi of the Phycomycetes, such as the common postharvest genera *Rhizopus* and *Mucor*, which contain many nuclei embedded within the cytoplasm and exhibit high radiation survival.

Radiation lesions in other components of the cell than DNA may also contribute to cell injury (Ichikawa 1981; Grecz et al. 1983), or may result in altered physiology which proves lethal (Sommer and Fortlage 1966).

Several radiation-induced morphological changes have been observed in fungi. These occurred mainly after spore germination and include changes in the dimensions of germ-tubes produced, the appearance of swellings in the mycelium, and the lack of cross-wall formation in species that normally form regular walls (Berk 1953; Sommer et al. 1964b; Sommer and Fortlage 1966).

The radiosensitivity of pathogens is determined by several factors:

Inherent Resistance. Response of fungi to inactivation (loss of colony-forming ability) by ionizing radiation is governed by several factors, of which the inherent resistance, which is genetically controlled, is the first (Sommer and Fortlage 1966; Alabastro et al. 1978; Moy 1983). Different fungal species may vary widely in their resistance to irradiation. For the same yeast species diploid cells were found to be more radiation-resistant than haploid cells (Latarjet and Ephrussi 1949), probably as a result of cell survival when one of the nuclei escapes inactivation by the applied radiation. A greater radioresistance has also been recorded for multinucleate spores than for unicellulate spores of the same fungus, suggesting that the escape of lethal injury by one nucleus would allow the spore to retain its colony-forming ability (Atwood and Pittenberg 1955). Several studies indicated that multicellular fungal spores, such as those of *Alternaria alternata* and *Stemphylium botryosum*, or bicellular spores, such as the mature pycnidiospores of *Diplodia natalensis* and many of the *Cladosporium herbarum* conidia population, are generally more resistant to gamma radiation than the unicellular spores of different fungal species (Sommer et al. 1964a,b; Barkai-Golan et al. 1968, 1969b). The sigmoid dose–response curves exhibited by *A. alternata* and *D. natalensis* are likely to be the consequence, at least in part, of the multicellularity and bicellularity, respectively, of their spores (Sommer et al. 1964b). The multicellular spore presumably gains resistance when one of the cells escapes injury and is capable of forming a colony.

The "black yeast", *Aureobasidium pullulans*, is another radioresistant fungus, whose importance considerably increases after irradiation of fruits and vegetables (Truelsen 1963; Skou 1964a). The high radioresistance of this microorganism was attributed to its polymorphism. In its older stages and under starvation conditions it produces radioresistant chlamydospores and a resting mycelium. The high radioresistance exhibited by young blastospores was, however, related to the natural formation of one or two buds in a large fraction of the blastospores, allowing them to behave as bi- or multicellular spores (Skou 1964a).

The problems posed in irradiation of multicellular structures of fungi, such as mycelia and sclerotia (Sommer et al. 1972; Paster and Barkai-Golan 1986), were mostly related to the ability of one cell in these structures to germinate and function independently of other cells (Sommer 1973).

158   Case Studies

Size of Population. The number of cells in the population, whether in the form of spores or mycelial cells, may greatly influence the radiation dose required to inactivate all or most of a population of identical cells. The difference in the initial number of cells in populations exposed to radiation was probably a reason for the wide range of "lethal doses" reported for *Botrytis cinerea* or *Rhizopus nigricans* (*R. stolonifer*) by different scientists (Beraha et al. 1959b, 1960; Skou 1960; Saravacos et al. 1962).

Working with *B. cinerea*-inoculated table grapes, Couey and Bramlage (1965) found that the effectiveness of a given dose was reduced with the increase in the density of spores in the inoculum. Furthermore, infection became increasingly resistant with age, probably due to the increased number of cells at a developed stage of infection. Beraha et al. (1961) found that a dose of 930 Gy significantly reduced young infections (1-day-old) of *Penicillium expansum* in Jonathan apples while a double dose was required to check decay in 4-day-old infections.

Barkai-Golan and Kahan (1966) showed that within the range of 1–2 kGy, the incubation period of the disease in *P. digitatum*- and *P. italicum*-inoculated oranges was gradually prolonged with the reduction in spore concentration. Along with the delay in fungal colonization the reduction in the initial spore concentration also resulted in a decreased rate of infection. Estimating the doses required to inactivate young cells or spores of prunus and citrus pathogens, Sommer et al. (1964a,b) pointed out the importance of considering both the inherent radiosensitivity of a fungus and the size of its population on the host, in determining the effective radiation doses. When the amount of fungal cells is large, even radiation-sensitive fungi may require relatively high doses for inactivation.

Rate of Radiation Application. For a given fungal species, the rate of inactivation increases with the dose. However, for a given dose, the rate of application may affect both spore survival and consequent mycelial growth. Beraha et al. (1964) found that *P. italicum*, irradiated at 1–2 kGy, formed fewer colonies on Czapek and Tochiani media after a high flux of 280 Gy/min than after a flux of 19 Gy/min. For the more resistant conidia of *B. cinerea* such an effect was found only on Czapek medium. Percent germination of conidia of the two fungi was influenced, however, only by the total dose delivered, independently of rate of application.

In *P. italicum*-inoculated oranges, a dose of 1370 Gy delivered at 400 Gy/min gave complete control of the blue mold rot for 12 days, whereas to achieve a similar rot control at 200 Gy/min a dose of 1820 Gy was necessary. No complete decay control could be achieved when a dose of 1820 Gy was delivered at 30 Gy/min (Beraha et al. 1964). Similarly, a higher flux enabled the reduction in the dose levels effective for control of *B. cinerea* in pears and *M. fructicola* in peaches (Beraha 1964), and resulted in better decay control in potato tubers inoculated with *Pythium debaryanum* (Beraha et al.

Radiation Effects on Pathogens

1959c). The greater radiobiological effect of a rapidly applied dose was suggested to result from the lack of, or the fewer, opportunities available for repair (Whiting 1960).

The Presence of Oxygen. Oxygen is a most important sensitizer and its presence in the atmosphere at the radiation site enhances the effectiveness of a given dose. Sommer et al. (1964c) found that the dose required to reduce survivors of various postharvest fungi to 1% in the presence of oxygen was only 60% of that in anoxia. The increased antimicrobial effect was attributed to the formation of injurious peroxides in the presence of oxygen (Laster 1954; Bridges and Horne 1959). Radiosensitization by oxygen, as proposed in the "damage fixation hypothesis" involves the formation of free radicals in the target molecule which then react with oxygen to form peroxy radicals.

Atmospheres of low oxygen tension, along with high CO_2 levels, may prevail around fruits sealed in selective plastic films, due to the continuous respiration of the fruit during storage (Barkai-Golan 1990). A modified atmosphere may also be created when fungi prepared for irradiation in vitro create near-anaerobic conditions by consuming the oxygen in a closed system (Sommer and Fortlage 1966). Under such conditions higher doses may be required for inactivation.

Another factor is the production of ozone following irradiation of oxygen. This gas may reach concentrations toxic to both host and pathogen, particularly when irradiation is performed in a closed container (Maxie and Abdel-Kader 1966).

Water Content. For both fungi and spore-bearing bacteria, higher sensitivity to radiation has been reported for vegetative cells than for spores. The high water content of bacterial vegetative cells may favor, within the cytoplasm, the production of a variety of harmful radicals which enhance radiation injury. These radicals are not as readily formed in the spores, which are dormant structures with little or no free water in their cytoplasm. In addition, the relatively high resistance of spores to ionizing radiation in aqueous suspension may, in part, be attributed to the impermeable multilayered coat which may create a barrier preventing the access of toxic chemicals. These may include radicals produced by radiation in water, such as the hydroxyl radical or H_2O_2, which elicit their effects through chemical interaction with cell constituents (Grecz et al. 1983).

Vegetative cells in the dehydrated or frozen state become markedly more resistant to ionizing radiation. Cellular water under these conditions is essentially immobilized and metabolism is arrested, and the radiation resistance of the recreative cells becomes comparable to that of spores (Grecz et al. 1983).

160   Case Studies

Host Protection. Microorganisms exhibit a greater resistance to irradiation within host tissues than under in vitro conditions. Such a phenomenon was recorded for *M. fructicola* conidia incubated after irradiation in sweet cherries and nectarines (Sommer et al. 1964a), or for *Penicillium cyclopium* and *P. viridicatum* conidia following inoculation in melon fruits (Barkai-Golan et al. 1968). No differences, however, were found between in vitro and in vivo rates for *Trichothecium roseum* and *A. alternata* conidia (Barkai Golan et al. 1968).

The lower radiation doses required for fungal inactivation within the host are believed to be the consequence of the chemical protective effect afforded by the tissues in contact with the spores (Sommer et al. 1964a).

Involvement of Phytoalexins. The ability of the plant tissue to synthesize fungitoxic compounds, phytoalexins, as a response to infection, may be an important factor in the resistance to infection in many host–parasite interactions (Kuc 1976). The accumulation of phytoalexins, however, is not dependent on infection and may also be caused by microbial metabolites, mechanical injury, various chemicals, irradiation, and other stresses.

Riov (1971) and Riov et al. (1971) reported the accumulation of the stress metabolites scopoletin (6-methoxy-7 dihydroxycoumarin) and scopolin (7-glucoside of scopoletin) in the peel of mature grapefruits irradiated at 1–4 kGy, as well as scoparone (6,7-dimethoxy coumarin) following irradiation at 3 and 4 kGy. Scoparone was absent from nonirradiated fruits whereas scopoletin and scopolin occurred in very low amounts, which increased gradually with the dose. These compounds were formed in the radiosensitive flavedo tissue of the peel in correlation with increased ethylene production, enhanced phenylalanine ammonia lyase activity, and the content of phenolic compounds which lead to cell death and peel pitting (Riov et al. 1968, 1972; Riov and Goren 1970; Riov 1975). No detection of scoparone was reported in March grapefruits following irradiation at 500–600 kGy (Moshonas and Shaw 1982).

Fifteen years after the first isolation of scoparone from irradiated grapefruits, following isolation of this compound from other citrus cultivars (Valencia oranges and Eureka lemons), this compound showed its antifungal activity (Dubery and Schabort 1987). Another irradiation-induced metabolite, extracted by Dubery et al. (1988) from damaged regions of citrus peel, was identified as 4-(3-methyl-2-butenoxy) isonitrosoacetophenone. This compound, which did not occur in extracts from nonirradiated fruits, was found to own antifungal capacity.

The formation of antifungal compounds in the host in response to radiation may thus be one factor in the complex of radiation effects on the pathogen, although its extent of importance in disease suppression has not yet been evaluated.

Radiation Effects on Pathogens   161

In contrast to radiation-induced phytoalexins in citrus fruit, El-Sayed (1978a–c) reported that the sesquiterpenoides rishitin and lubimin formed in potato tubers and rishitin formed in tomato fruits in response to infection, were significantly reduced after gamma irradiation. In parallel with this reduction an increase in the incidence of rotted tubers and fruits was frequently reported after exposure to radiation doses required for potato tuber sprout inhibition (100 Gy) and for tomato fruit shelf-life extension (3 kGy) (Duncan et al. 1959; Hooker and Duncan 1959; Cotter and Sawyer 1961; Mathur 1963; Truelson 1963; Abdel-Kader et al. 1968). The increase in the incidence of decay was suggested to result from the lowered capacity of the irradiated tissue to form antifungal compounds in response to infection. Furthermore, postirradiation treatment of potato tubers and tomato fruits with phytoalexins, produced by the same plant organ or by another member of the Solanaceae, increased the efficiency of irradiation in controlling microbial spoilage (El-Sayed 1978c).

Timing of Irradiation. Extending the time lag between inoculation and irradiation may increase dose requirements for pathogen suppression (Zegota 1988). Working with *M. fructicola*-inoculated peaches, Kuhn et al. (1968) found that extending the interval time between inoculation and irradiation from 24 to 36, 48, or 60 h increased the incidence of lesions from 10 to 60, 80, or 90%, respectively. Spalding and Reeder (1986a) found that reduction in bacterial soft-rot in mature, green tomatoes inoculated with *Erwinia* sp. was greatest when the fruit was irradiated at 1 kGy within 3 h after inoculation. Since many of the postharvest pathogens are "wound parasites" which gain entrance into the fruit or vegetable through unavoidable wounds occurring at harvest, the time lag between harvest and irradiation would parallel the time between inoculation and irradiation. During this period, the infection may be initiated and the size of population exposed to irradiation may increase.

For latent fungi, an extension of the time interval between harvest and treatment may result in renewal of fungal development due to the progress in the ripening process of the fruit (Prusky et al. 1982; Droby et al. 1986).

Pre- and Postirradiation Conditions. Since the fungicidal or fungistatic effect of irradiation is associated with the size of the target population, the presence of environmental factors favorable for fungal growth will indirectly affect dose requirements.

The storage temperature is a major environmental factor which may determine the rate of spore germination and consequent fungal growth. Storage temperatures of 0 °C arrest growth of most postharvest fungi except for a few cold-tolerant species (Muller 1956). However, many crops, mainly tropical and subtropical ones, are cold sensitive and have to be stored at higher temperatures which favor growth of most pathogens. Even the

162   Case Studies

natural temperature prevailing in the orchard may be responsible for the degree of infection by pathogens that penetrate the host prior to harvest (Shiffmann-Nadel and Cohen 1966).

High relative humidity or high moisture content in the air, which are favorable for spore germination, are other factors that may determine the rate of colonization. Such conditions are regularly applied for many commodities to prevent dehydration during storage. Controlled atmosphere storage with an elevated CO_2 level and/or reduced O_2 tension, as well as a modified atmosphere formed around the produce within selective polymeric film packages, may also suppress pathogen growth either directly or via increasing host resistance (Barkai-Golan 1990).

Formation of Mutants. Exposure of microorganisms to sublethal radiation doses may induce various degrees of injury and result in the formation of mutants among the survivors. Of particular concern regarding fruits and vegetables is the possible appearance of mutants which will differ from their parents by a higher resistance to radiation and by increased pathogenicity to the harvested crop (Sommer and Fortlage 1966).

Although mutants usually appear to be less pathogenic and less vigorous than the parent organism (Beraha et al. 1964), mutants of wider virulence were also observed in different plant pathogens (Day 1957; Flor 1958). A white, sporeless mutant of *P. digitatum*, obtained by Barkai-Golan et al. (1966) after irradiation at 1910 Gy, was found to be considerably more sensitive to further irradiation than the parent strain. The mutant retained its pathogenicity to citrus fruits, and exhibited cellulolytic activity similar to that of the original fungus, but its rate of development at suboptimal temperatures (8–17 °C) was slower (Barkai-Golan et al. 1966; Barkai-Golan and Karadavid 1991).

Damage Repair Capacity. The ability of a pathogen to repair its radiation damage under postirradiation conditions may determine the dose requirements for treatment (Sobels 1963; Sommer and Fortlage 1966).

The cell is equipped with replication and repair enzymes which may carry out either DNA synthesis or, alternatively, DNA degradation. The DNA–enzyme assembly seems to exist in a precariously balanced state and whether synthesis or degradation will follow initial radiation injury seems to be controlled by a multitude of factors which are still poorly understood (Grecz et al. 1983). Although the inherent mechanisms of radiation-induced DNA damages are not entirely clear, there are indications of rejoining repairs specific to ionizing radiation (Ichikawa 1981).

In vitro studies with various microorganisms suggest that encouraging environments for recovery processes are characterized by their ability to slow down metabolism, such as incubation of treated organism at suboptimal temperatures or growing under starvation or anaerobic conditions

Radiation Effects on Pathogens

(Stapleton et al. 1953; Latarjet 1954; Alper and Gillies 1958; Sommer et al. 1963a, 1964c, 1965; Sommer and Creasy 1964). Studies with *R. stolonifer* sporangiospores irradiated at potentially lethal doses (5 kGy) indicated that some restoration to the nonlethal condition could be achieved by delaying germination (Sommer et al. 1963a, 1964c; Buckley et al. 1967). When germination was inhibited by the presence of an unsuitable medium or by anaerobiosis, the number of repaired spores was greatest at a temperature near the optimum for growth (Sommer et al. 1963a). Spores of *R. stolonifer* sensitized to irradiation by iodoacetamide were found to retain a considerable capacity to recover. However, the size of the spore fraction capable of recovery was reduced as a consequence of sensitization. This effect on recovery might be an indirect effect on spore metabolism rather than a direct interference with recovery mechanisms (Sommer et al. 1971).

Studies of the direct effects of ionizing radiation on pathogen development have frequently been conducted as the first step in evaluating the potential for postharvest disease control. Spores are considered to be the most convenient fungal structure available for dose–response studies. Being more resistant to radiation than the vegetative cells, spores are frequently preferred for evaluating the efficiency of irradiation on disease suppression. There are, however, a few drawbacks in using spores for dose–response studies. Moy (1983) listed them as follows:

1. a sufficient number of spores is not always available;
2. some of the large spores are of indefinite multicellularity and are difficult to quantify on a cellular basis;
3. sometimes more than one type of spore may be present in a culture and a uniform suspension cannot be readily obtained; and
4. in many instances it is the mycelia rather than the spores that are inactivated in a fruit.

The difference in fungal response to radiation was generally exhibited in vitro by "spore survival", i.e., the ability of fungal spores to form colonies. This feature was found to be more sensitive to radiation than the germination ability of spores (Beraha et al. 1959a,b; Nelson et al. 1959). The dose–response curves for spores of common postharvest pathogens emphasized the wide range in their radiosensitivity. For each species, however, the rate of spore inactivation increases with the dose (Fig. 25).

The doses required to inactivate a large percent of fungal populations can be performed by "end point" technique. Sommer et al. (1964a,b) studied the doses required to inactivate every cell in populations varying from 10 to 10^6 young fungal cells or spores of postharvest pathogens of prune and citrus fruits. Percents of populations that failed to grow were plotted against dose and various inactivation levels were approximated by interpolation. In most cases the dose required to inactivate every cell within 80% of fungal populations of a particular size was established.

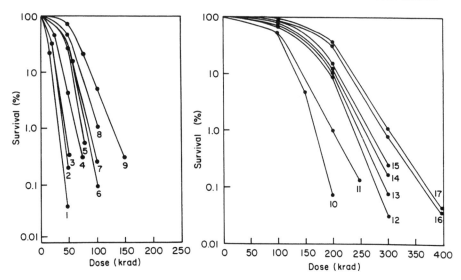

Fig. 25. Dose-response curves for inactivation of postharvest fungi by γ-radiation. *1 Trichothecium roseum*; *2 Trichoderma viride*; *3 Phomopsis citri*; *4 Penicillium italicum*; *5 P. expansumi*; *6 Asperigillus niger*; *7 P. digitatum*; *8 Geotrichum candidum*; *9 Monilinia fructicola*; *10 Botrytis cinerea*; *11 Diplodia natalensis*; *12 Stemphylium botryosum*; *13 Rhizopus stolonifer*; *14 Alternaria citri*; *15 A. alternata*; *16 Cladosporium herbarum*; *17 Candida* sp. (Sommer et al. 1964a; Barkai-Golan and Kahan 1971)

The relative resistance of a pathogen to radiation can be expressed by the lethal dose required for permanent growth inhibition (Barkai-Golan et al. 1968). However, irradiation can also be used at a fungistatic level which is capable of temporarily halting the growth. The time lag before growth is resumed, as well as the rate of subsequent mycelium growth, have been commonly used as parameters for fungal response to radiation (Beraha et al. 1959a,b; Sommer et al. 1964a; Barkai-Golan et al. 1967, 1968; Mahmood 1972). Similar to the length of the time lag in vitro, sublethal doses can also extend the incubation period of the fruit disease (Beraha et al. 1959a,b; Nelson et al. 1959; Sommer et al. 1964a; Barkai-Golan et al. 1966, 1967, 1968, 1969b). The ability of sublethal doses to delay by several days the onset of fungal colonization may be of advantage for fruits, such as strawberries and cherries, that are characterized by a short postharvest life and valuable additional lesion-free days.

Working with a *R. stolonifer* culture, Sommer et al. (1963b) found that the ability to produce cell wall-degrading pectolytic enzymes was more radiation resistant than the potential for colony formation or the ability of sporangiospores to germinate. Spores that became incapable of forming colonies after irradiation continued to produce pectin glycosidase for 6 days

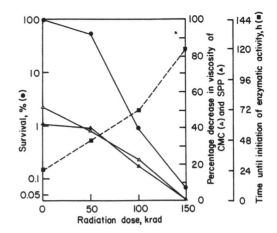

Fig. 26. Survival of *Penicillium digitatum* spores after irradiation (●) in relation to time until initiation of enzymatic activity (■) and activity of cellulase (△) and polygalacturonase (▲) as expressed by percentage decrease in viscosity of carboxymethylcellulose (CMC) and sodium polypectate (SPP), respectively, after 3 days on potato dextrose agar. (Barkai-Golan 1971)

regardless of the amount of radiation the spores had received, unless the dose exceeded about 10 kGy. The effect of radiation on the pectolytic ability of the pathogen was related to the dose affecting the swelling of the spore prior to germ-tube formation. Irradiation of *P. digitatum* conidia with sublethal doses (Barkai-Golan 1971) resulted in a dose-dependent lag in the polygalacturonase and cellulase activities of the fungus. A lag of 24 h was obtained with 500 Gy and a lag of 5 days with 1.5 kGy. The inhibition of enzymatic synthesis in an irradiated conidium population was found to be directly related to the number of conidia which remained viable after corresponding doses of gamma rays (Fig. 26).

In addition to the direct destruction of the pathogen, there is also a possibility of indirect suppression via ripening inhibition. Radiation doses required for direct suppression of postharvest pathogens are generally above the tolerance level of the fruit and result in radiation-induced damage. However, for several fruits, a reduction in the incidence of fungal diseases has been recorded after exposure to relatively low radiation doses which are incapable of directly suppressing or even temporarily retarding pathogen growth, but which are sufficient to delay the ripening and senescence processes. Postharvest infection is greatly enhanced with fruit maturity (Eckert 1978). Several factors may be involved in the increased resistance to infection of the unripe fruit, such as the lack of suitable nutrition, the accumulation of toxic compounds in the immature tissues or the failure of the fungi to synthesize and activate cell wall-degrading enzymes in the immature tissue (Eckert 1978). Furthermore, the ripe fruit is more vulner-

166   Case Studies

able to injury and thereby more exposed to infection by wound pathogens (Eckert and Sommer 1967). Factors that delay ripening will thus indirectly contribute to disease suppression by maintaining host resistance to infection.

Extensive research has clarified that irradiation doses in the range of 50–850 Gy may inhibit ripening of mango, papaya, banana and other tropical and subtropical fruits (Akamine and Moy 1983; Thomas 1985). The ability of low doses to alter ripening depends not only on fruit species but also on cultivars of the same fruit. Other factors, such as the stage of maturity at the time of treatment, time of irradiation after picking, and storage conditions may all influence host response to ripening inhibition.

Alabastro et al. (1978) found that irradiation of mature green mangos of the carabao cultivar with 160, 180 and 220 Gy doses delayed the appearance of anthracnose and stem-end rot by 3 to 6 days without adverse effect on fruit appearance. The suppression of decay by such low doses suggests that no direct fungicidal effect on the pathogens is involved. Furthermore, the lower doses were more effective in decay suppression than the 220 Gy dose, suggesting that the resistance of the fruit to storage infection has been reduced by the absorbance of the higher dose. Irradiating freshly harvested Hawaiian Solo papayas, Akamine and Wong (1965) and Akamine and Goo (1977a) found that a dose of 750 Gy delayed ripening by more than 2 days and the rate of overripening (senescence) by 3 days at room temperature. The delay of ripening and senescence was not increased at 1 kGy. Irradiation at 250–750 Gy was found by Jiravatana et al. (1970) and Loaharanu (1971) to result in up to 7 and 10 days delay in the ripening of Puerto Rican and Thai papayas, respectively, when applied at the preclimacteric stage. In parallel, doses of 250–500 Gy also resulted in the reduction of decay control. Since, for a direct suppressive effect on the pathogens, higher doses would be required, decay suppression at this dosage seems to be primarily related to the delay in ripening. It was, however, noted that decay reduction at the ripening retardation doses was only marginal while doses of 1 kGy and more are above the tolerance level of the fruit (Moy et al. 1973; Moy 1977a,b). Doses of 50–370 Gy were sufficient to significantly reduce decay by *Colletoricum* but not by *Thielaviopsis* in Phillipines Cavendish bananas (Alabastro et al. 1978). The reduction in anthracnose by such low doses was probably the result of ripening inhibition. However, doses of 100–800 Gy caused darkening of the peel which increased with the dose.

Prusky et al. (1982, 1983) isolated from the peel of unripe avocado fruits a preformed antifungal diene compound (cis, cis-1-acetoxy-2-hydroxy-4-oxo-heneicosa-12,15-diene) which was suggested to account for latency of *C. gloeosporioides* infection in the unripe fruit. This compound, which inhibits vegetative growth and spore germination of *C. gloeosporioides* at concentrations found in the unripe fruit peel, decreases during ripening to subtoxic levels. Irradiation of Fuerte avocados at 50 and 200 Gy doses was found to enhance the level of the diene compound immediately after treat-

Radiation Effects on Disease Development 167

ment (Prusky, pers. commun.). However, 1 day after treatment the level of the antifungal compound had already decreased and by the 3rd day its level in the irradiated fruit was considerably lower than in the untreated control fruit. It was suggested that irradiation of avocado induces the formation of this diene compound for only a very short time which is then followed by enhanced fruit ripening along with increased susceptibility to decay.

5.1.3 Radiation Effects on Disease Development

Determining the radiation dose required for pathogen inactivation in vitro may serve in some cases as a basis for evaluating the dose required for disease suppression. However, pathogen response within the host may differ from that in culture because of the various host constituents met by the invader and of other host–pathogen interactions. These include the inducement of antifungal phytoalexins as a response to infection (Kuc 1976) or the production of ethylene during pathogenesis. Ethylene may affect pathogens either directly by stimulating growth or indirectly by enhancing ripening and scenescence processes (El-Kazzaz et al. 1983; Barmore and Brown 1985; Barkai-Golan et al. 1989). Such effects may change the size of pathogen population exposed to radiation.

The inactivation dose for microorganisms is generally higher when they are located within the tissues, possibly due to the protection afforded by host constituents. In addition, suppressive radiation doses may differ according to the actual population size of infecting microorganisms at the time of irradiation. This size, which may vary within a commodity, is difficult to estimate.

The factor which practically determines the radiation dose for disease control is the tolerance of the host to radiation rather than the fungicidal dose required for pathogen suppression. Different host species, and even different cultivars of a given species, may differ in their tolerance to radiation. Furthermore, dose tolerance may be influenced by the stage of fruit ripeness at the time of treatment and by the subsequent conditions and duration of storage (Maxie and Abdel-Kader 1966).

Studies undertaken in the 1950s and 1960s have already clarified that fruits and vegetables are usually susceptible to radiation doses that are lethal to the common postharvest pathogens. However, the same studies have also indicated that extension of postharvest life of several commodities can be achieved by using sublethal doses, which can inhibit fungal growth temporarily and thus prolong the incubation period of the disease (Beraha et al. 1959a,b; Beraha et al. 1961; Beraha 1964; Sommer et al. 1964a,b; Kahan and Monselise 1965; Barkai-Golan and Kahan 1966). Such a prolongation may be sufficient for decay control in fruits with a short postharvest life and for which several additional days free of lesion are valuable.

168   Case Studies

Strawberry. This is an example of a fruit for which sublethal doses are sufficient to prevent decay for several days without causing damage to the fruit. For a fruit characterized by a short postharvest life on the one hand and by a high market value on the other hand, a few days' delay in deterioration may be very beneficial. As a matter of fact, strawberry has frequently featured as the first product in the list of fruits and vegetables that benefit from radiation as a fungistatic treatment (Sommer and Fortlage 1966; Sommer and Maxie 1966; Dennison and Ahmed 1975; Moy 1983; Barkai-Golan 1985).

Postharvest life of strawberries is limited mainly by the development of the grey mold, caused by *B. cinerea*, and the watery decay, caused by *R. stolonifer*, although other Mucoraceae may also be associated with postharvest spoilage (Sommer and Fortlage 1966; Dennis 1983). Preharvest fungicidal sprays considerably reduce the incidence of the grey mold fungus, which can normally penetrate the fruit in the field (Powelson 1966). This treatment cannot, however, by itself supply adequate control (Aharoni and Barkai-Golan 1987). During cold storage or refrigerated transit, *Botrytis* growth is retarded while *Rhizopus* growth is totally inhibited. However, normal growth resumes at ambient temperatures.

Beraha et al. (1961) found that within the range 930–2790 Gy an appreciable reduction in decay, by both *B. cinerea* and *R. stolonifer*, was recorded in strawberries after 3 days' storage at 24 °C or 10 days' storage at 5 °C. When the fruit was stored for 2 days at 24 °C, followed by 4 days at 5 °C, no rot had developed at any of these doses while the nonirradiated controls were almost completely rotted. Adverse effects, such as softening, leaking, and bleaching, were noted only at doses of 4650 Gy. Although a dose of 2 kGy was defined by Cooper and Salunkhe (1963) as optimal for preservation of strawberries, it was emphasized that different cultivars responded differently to irradiation and that the cultivars Kasuga, Lindalicious and Sparkle, when irradiated at the firm-ripe stage, kept better than Marshall, Robinson or Shasta strawberries. Under actual and simulated marketing conditions Maxie et al. (1964b) found that irradiation at 2 kGy significantly decreased decay in Shasta strawberries without causing any apparent ill effects, provided that the berries were cooled immediately after harvest and were kept cool. This dose did not affect the ascorbic acid content of the berries. Loss of aroma and flavor were recorded only after 21 days in cold storage. Irradiating naturally infected Lassen strawberries with a dose of 2 kGy, Barkai-Golan et al. (1971) found that the incubation period of the grey mold was extended from 3 to 10 days at 15 °C and that the subsequent rate of growth of the surviving fungal population was markedly reduced (Fig. 27). Such a delay in the initiation of *Botrytis* infection may also indirectly reduce the level of contact infection and nesting during storage.

A dose of 2 kGy was also effective in improving the keeping quality of Jugoslavian strawberries (cv. Pocahontas) (Bogunowic et al. 1986), whereas

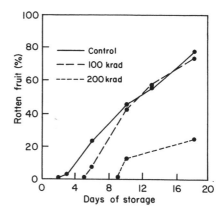

Fig. 27. Effect of gamma radiation on the incidence of decay and the incubation period in Lassen strawberries held at 15°C. (Barkai-Golan et al. 1971)

doses of 2.5 or 3 kGy applied 6–10 or 20–24 h after harvesting were required to extend the cold storage of Dukat strawberries (Zegota 1988). Better results were reported for fruit with stems and when the time between harvest and irradiation was short. Irradiation did not change the titratable acidity and content of reducing sugar, whereas ascorbic acid level and color intensity decreased with the absorbed dose and storage duration. Under conditions of fully commercial application of radurization in South Africa, Du Venage (1985) found that the summer shelf-life of strawberries has been extended from an average of 3–12 days at ambient temperatures up to 50 days at 2°C.

Grapes. The possibility of suppressing *B. cinerea* and other postharvest pathogens of various cultivars of table grapes by irradiation has also been considered. Beraha et al. (1961) found that the time of appearance and intensity of decay in *B. cinerea*-inoculated Tokay grapes were inversely proportional to the dose. After exposure to 1860 Gy the berries remained free of decay for 4 days at 24°C whereas to prevent decay for 10 days' storage, a dose of 4650 Gy was required. Increase in postharvest life was reported by Rogachev (1966) for several grape cultivars at doses of 2–3 kGy. Storage duration was extended to 15–30 days at 25°C compared to 4–7 days for the nonirradiated control. Doses above 4 kGy caused yellowing in some cultivars and reduced fruit resistance to repeated infections. Working with Hoanes grapes, Belli-Donini and Pansolli (1970) found that irradiation at 2 kGy delayed both the onset of fungal attack and the incidence of decayed berries. This treatment resulted in the extension of storage life by 18 days not only from pathological aspects but also from physiological aspects or changes in color, flavor, and texture. Doses of 1 and 2 kGy retained the marketability of Greek grape cultivars packed in perforated polyethylene bags for up to 60 and 80 days at 0°C compared to 35 days for the nonirradiated fruit (Saravacos and Macris 1963). Spoilage in irradiated

170 Case Studies

grapes was caused mainly by *Pullularia pullulans*, which is exceptionally resistant to radiation (Saravacos et al. 1962; Saravacos and Macris 1963). No changes in color or firmness of the berries in irradiated Tokay grapes were reported by Beraha et al. (1961), and in irradiated Thompson Seedless grapes by Salunkhe (1961). However, several studies have emphasized the adverse effects of irradiation on grapes. Maxie et al. (1964a), studying the response of Emperor and Tokay grapes, Bramlage and Couey (1965), working with Emperor and Thompson grapes, and Barkai-Golan and Kahan (1968), working with Thompson Seedless and Alphonse Lavallée cultivars, found that irradiation at 2–3 kGy resulted in loss of firmness and impairment of fruit quality, and was regarded as in effective for fruit preservation and impractical as a substitute for SO_2 treatment.

Stone Fruits. The major postharvest diseases of peaches, apricots, nectarines, plums, and cherries are the brown rot caused by *M. fructicola*, the watery rot caused by *R. stolonifer*, and the grey mold caused by *B. cinerea*. Since *Monilinia* growth may be halted at temperatures below 5°C and *Rhizopus* growth is inhibited even at temperatures below 10°C, adequate refrigeration may provide good decay control for these pathogens. However, when the fruit is removed from cold storage for final ripening, fungal growth is resumed (Sommer and Maxie 1966). Studying the in vitro response of *M. fructicola* to gamma radiation, Sommer et al. (1964) found that inactivation dose increased from 2 to 3 kGy as the population density was raised from 10^4 to 10^6 spores/ml. The dose needed for controlling brown rot infections on peaches was found by Beraha (1964) and Beraha et al. (1959b) to be 1250–1500 Gy in a high dose-rate radiation source (250 Gy/min), but more than 2 kGy in a low dose-rate source (25 Gy/min). The time interval between inoculation and irradiation was another factor affecting the dose needed for fungal control (Kuhn et al. 1968). When "firm-ripe" Dixiland peaches were irradiated at 2 kGy within 24 h after inoculation, only 10% of the inoculations formed lesions. Postponing irradiation to 36, 48, and 60 h after inoculation increased the incidence of lesions to 60, 80, and 90% respectively. Rogachev (1966) reported that although ripe peaches spoiled more rapidly than green peaches, doses of 3 and 4 kGy were required for shelf-life extension of both ripe and green fruits. Working with plums and nectarines, Brodrick et al. (1985) observed that although doses of least 2 kGy controlled the development of *R. stolonifer*, *M. fructicola*, and *B. cinerea* this treatment caused a slight change in host texture and resulted in unmarketable products. In an earlier study with nectarines and peaches Maxie et al. (1966) found that rates of CO_2 and ethylene production in fruits subjected to 1–6 kGy were much higher in nonirradiated than in irradiated fruits at every stage of the climacteric sequence, and that irradiated fruits ripened 4–5 days earlier. Irradiated nectarines were acceptable for flavor and appearance but were characterized by a loss of flesh

Radiation Effects on Disease Development

firmness which made them so susceptible to mechanical injury that irradiation as a protective technology was precluded. Working with sweet cherries, Cooper and Salunkhe (1963) found that a dose of 3 kGy extended the refrigerated life of these fruit beyond 30 days. This dose caused a total inhibition of *Penicillium* decay, whereas the radioresistant fungi *Alternaria* and *Stemphylium* predominated on the irradiated fruit. At 4 kGy the cherries turned brown while pigment development and ripening were progressively retarded. Massey et al. (1965) found that irradiation at 2 kGy and higher doses drastically reduced the microbial contamination on Napoleon cherries. However, in spite of the initial substantial reduction, a higher percent of moldy fruit was recorded when the storage period was extended. The inability of higher doses to prolong the storage life of cherries free from brown rot seemed to be the result of increased susceptibility of the host to invasion by spoilage organisms that had survived the treatment. Softening, which had already been detected at 500 Gy, progressed with the dose and was related directly to the degradation of pectic constituents of the tissues.

Pome Fruits. Evaluating the relative radiosensitivity of four postharvest pathogens of pears, Barkai-Golan et al. (1969b) showed that *P. expansum* and *B. cinerea* were radiation sensitive, whereas *Alternaria tenuis* (*A. alternata*) and *S. botryosum* were both radiation resistant. For each species, the rate of germination inhibition increased with the dose from 0.5 to 3 kGy and decreased with spore concentration subjected to irradiation. In vitro studies by Beraha et al. (1961) indicated that germination of irradiated conidia of *P. expansum* and *B. cinerea* was dependent both on the medium and the length of time for which the conidia were held after irradiation and before being plated out. Doses between 2790 and 4650 Gy were required to prevent *P. expansum* germination in both Czapek and Tochiani media, but rate of recovery was higher in Tochiani medium. Considerably higher doses were, however, required to prevent germination of *B. cinerea*. Working with *P. expansum*-inoculated Jonathan apples, Beraha et al. (1961) found that a dose of at least 930 Gy was required for significant reduction in rot development after 6 days' storage at 24°C, when the fruit was inoculated 1 day prior to irradiation. After 10 days' storage under these conditions, a double dose was required to reduce the advance of infection. However, based on the assumption that *Penicillium* lesions were not present on the fruit at harvest, much lower doses were suggested to be effective in controlling blue mold disease. Some prolongation of the incubation period of both *P. expansum* and *B. cinerea* in Spadona pears occurred after irradiation at 1 kGy (Barkai-Golan et al. 1969b). With the increase in dosage up to 3 kGy the incubation period was further prolonged and the percent of infected fruit started to decrease after a dose of 2 kGy. For the two resistant species, *S. botryosum* and *A. alternata*, a decrease in germination ability

172 Case Studies

occurred at 2–3 kGy, but pathogenicity to pears and length of the incubation period were not significantly affected. However, the main limitation of irradiation in both apples and pears was the radiation-induced softening and skin damage at effective doses (Beraha et al. 1961; Barkai-Golan et al. 1969b). Irradiating three Chinese apple cultivars, Zhao and Wan (1987) found that both respiration and ethylene production increased when irradiation took place shortly after picking. Since, at a dose of 800 Gy applied 10 days after picking, the respiration rate returned to normal within 5 days and ethylene production decreased to a very low level, this dose was recommended for fruit preservation.

Citrus Fruits. Postharvest citrus pathogens are characterized by a broad range of radiosensitivity: from the most sensitive species, *Trichoderma viride* and *Phomopsis citri*, through the relatively sensitive *P. italicum*, *P. digitatum*, and *Geotrichum candidum* to the radioresistant *D. natalensis* and *A. citri* (Fig. 25). Studies by Beraha et al. in the late 1950s (1959a) showed that irradiation of citrus fruit inoculated with *P. digitatum* at 1410–1860 Gy delayed rotting by 17 days at 13 °C. The extension of the incubation period of the disease in *P. digitatum*- and *P. italicum*-inoculated fruits was found to increase with the radiation dose but to decrease with the initial size of spore population (Sommer et al. 1964b; Barkai-Golan and Kahan 1966). At 10^2 spores per inoculum, decay development was entirely prevented by 2 kGy during 80 days at 23 °C (Barkai-Golan and Kahan 1966). A linear relationship between the extension of fruit shelf-life and radiation dose at the range of 450–1750 Gy was found in Valencia oranges irradiated 4 h after inoculation with *P. digitatum*. With each increment of 141 Gy an additional day of storage at 4–60 °C was achieved (Kahan and Monselise 1965). Mahmood (1972) found that doses of 2–3 kGy controlled effectively postharvest decay of citrus fruits caused by *P. italicum*, *P. digitatum*, *Phytophtora citrophtora*, *Phytophthora parasitica*, and *G. candidum*. However, a 2 kGy dose already impaired the organoleptic quality of the fruit.

Several studies evaluated the effects of superficial electron irradiation on pathogenic fungi which infest the peel of citrus fruits (Kahan et al. 1968a; Umeda et al. 1969; Ojima et al. 1973, 1974, 1975; Chachin et al. 1974; Watanabe et al. 1976). Irradiating Valencia oranges, inoculated superficially with *P. digitatum* spores, with a 520 keV electron beam, Barkai-Golan and Padova (1981) reported that the incidence of green mold rot decreased with increasing dose from 0.5 to 1 kGy, but a further increase to 1.5 kGy did not improve the results.

It was generally found that the peel of citrus fruits is very sensitive to gamma or electron radiation at doses required for decay control, or even lower (Grierson and Dennison 1965; Dennison and Ahmed 1966, 1971; Kahan and Padova 1968; Barkai-Golan and Padova 1981). On the other hand, doses suitable for controlling the two penicillia and other relatively

Radiation Effects on Disease Development

sensitive pathogens cannot protect citrus fruits against *A. citri* (Dennison and Ahmed 1971) and frequently there is a high incidence of stem-end rot caused by this fungus. The extent of damage was significantly reduced when the fruit was coated with a wax layer or wrapped in polyethylene film prior to irradiation (Barkai Golan and Padova 1981). However, experiments carried out by Ojima et al. (1973) with *Citrus unshiu* showed that polyethylene, as well as other plastic films, were unsuitable for long-term storage due to the formation of conditions favorable for fungal development.

Avocado. The avocado is extremely sensitive to radiation and various cultivars exhibit phytotoxic effects at doses as low as 50–100 Gy. Irradiation studies with avocados were therefore focused on the possible use of low doses for delaying fruit ripening and senescence and as a physical quarantine treatment for disinfestation of fruit flies. The doses optimal for delay of ripening vary widely with fruit cultivar. For a given cultivar this dose may differ with the area of growing, the stage of fruit maturity, and the date of picking during the season. Irradiating four Isreal avocado cultivars (Ethinger, Fuerte, Nabel, and Hass), Kahan et al. (1968a) found marked differences in ripening, ranging from substantial delay in Ethinger to acceleration in Hass avocados. For Ethinger fruits, irradiation was most effective in fruit picked early in the harvest season, when a delay of 24 days was recorded after a 350 Gy dose. However, not only was no effect on decay development recorded, but the irradiation resulted in peel damage which increased when applied to wet fruits or to fruits predisposed to cooling at 6 °C. Several studies reported a few days' delay in ripening of Fuerte avocados after irradiation at 25–100 Gy (Young 1965; Kamali et al. 1972; van der Linde 1982). On the other hand, some acceleration accompanied by browning of the vascular bundles was noted by Bramlage and Couey (1965) in Fuerte avocados irradiated at 100 Gy.

The high radiosensitivity of avocados renders irradiation as ineffective for decay control. For improved postharvest life along with reduced decay, combined treatments of dipping the fruit in hot water and exposing them to very low radiation doses have been developed (vide infra). The most likely application of irradiation for avocados is its use as an alternative quarantine treatment for disinfestation of fruit flies, where doses as low as 25 Gy may suffice to prevent the production of viable fruit fly offspring (Moy 1983; Thomas 1986a).

Mango. Studying the effects of irradiation in the range 150–1500 Gy on decay and quality of Florida mango cultivars, Spalding and Windeguth (1988) found that severity of anthracnose caused by *C. gloeosporioides* was reduced by exposure to 500 Gy or more in Keitt mangos but not in the Tommy Atkins variety. The severity of stem-end rot caused by *D. natalensis* or *P. citri* in Tommy Atkins mangos was reduced by a dose of 1.5 kGy,

174   Case Studies

whereas a 750 Gy dose was not effective. These data are in accord with reports by Ahmed et al. (1968) and Akamine and Moy (1983) which showed that dosages higher than 1 kGy were required for substantial reduction of fungal decay in mangos. In Kensington Pride mangos, Johnson et al. (1990) found that doses ranging from 300 to 1200 Gy reduced anthracnose (*C. gloeosporioides*) and stem-end rot (*Dothiorella dominicana*) at 20 °C but the level of control was not commercially acceptable. Brodrick et al. (1972), working with South African mangos (Kent, Peach, and Sabre cultivars), found that control of anthracnose by irradiation was inconclusive, while the development of diplodia stem-end rot could not be successfully controlled even at doses of up to 2 kGy. However, studies with different mango cultivars showed that low irradiation doses, which are ineffective in direct suppression of fungal growth, can be beneficial in extending the normal shelf-life of the fruit by slowing down the rate of ripening and senescence processes (Mathur and Lewis 1961; Dharkar et al. 1966a,b; Mumtaz et al. 1970; Nair et al. 1973; Farooqui et al. 1974a; Heins 1977; Rashid and Farooqui 1984; Lunt 1985; Boag et al. 1990), and thus maintaining fruit resistance. Postharvest irradiation of mangos at dose levels effective in delaying the processes of ripening and senescence were also found to be effective as a quarantine treatment against fruit flies and mango stone weevil infestation (Thomas 1985).

Papaya. Several studies have indicated the possibility of using low doses of radiation as a means for delaying the rate of ripening and senescence in papayas and as a disinfestation treatment for quarantine purposes against fruit flies. Moy et al. (1973) and Moy (1977a) showed that irradiating papayas grown in Hawaii, Taiwan, and Venezuela at low doses of radiation, aimed at fruit fly disinfestation and shelf-life extension, was effective. However, anthracnose, which is generally considered to be the major postharvest problem, could not be controlled at these doses. Doses of 750–1000 Gy, which delayed ripening and senescence by 2–3 days, only slightly suppressed storage decay in Hawaiian Solo papayas (Akamine and Wong 1965; Akamine and Goo 1977a). On the other hand, doses of 250–500 Gy, which delayed ripening for up to 7 days in mature, green Puerto Rican papayas, also resulted in decay control when the fruit was irradiated 1 day after harvest (Jiravatana et al. 1970). The maximum dose that could be tolerated by fruits in the green preclimacteric or color-turning stage was found to be 1 kGy and above this level fruits developed surface darkening (Akamine and Wong 1965). A tolerance level of 1 kGy was also found by Moy et al. (1973) for mature, green to partially ripe fruit of Hawaiian and Taiwanese papayas, and a tolerance level of 1–1.5 kGy was reported for Venezuelan papayas. In order to improve the effect of low doses for fly disinfestation, delay of fruit ripening, and control of storage decay, combined heat and radiation treatments have been developed.

Radiation Effects on Disease Development 175

Banana. In vitro studies by Saravacos et al. (1962) clarified that most postharvest pathogens of banana are radiation resistant. Studying the response to radiation of fungal pathogens of Lacatan bananas in the Philippines, it was found that the lethal dose for *Gloeosporium musarum* and *C. gloeosporioides* was 4 kGy, whereas lethal doses for *Botryodiplodia theobromae* and *Pestalotia* sp. were 5 and 3 kGy respectively. Working with bananas inoculated with *Gloeosporium musarum* spores (1.5×10^5 spores/ ml) Ferguson and Yates (1966) found a 50% survival of the fungus after a 1 kGy treatment. Anthracnose development was markedly ratarded by treatment with 1.5 kGy whereas doses of 2.5–3 kGy resulted in complete growth inhibition (Meredith 1960; Ferguson and Yates 1966). However, several studies emphasized the high sensitivity of bananas to irradiation, indicating that much lower doses may result in radiation damage to the fruit (Kahan et al. 1968b; Alabastro et al. 1978).

Following an early report by Brownell (1952) that low doses of 93 Gy are capable of delaying ripening in bananas, most irradiation studies have been focused on the extension of postharvest life of the fruit by retarding the ripening process. The extent of ripening delay was found to differ with fruit cultivar: while no delay was found by Dharkar and Sreenivasan (1966) following irradiation of some Indian cultivars, doses up to 0.4 kGy delayed ripening to a different extent in some Israeli Dwarf Cavendish bananas irradiated 1–7 days after picking (Kahan et al. 1968a). Reports from the United States by Maxie et al. (1969) and Chowdhury and Hamid (1969) indicated that doses up to 0.5 kGy inhibited ripening without harming fruit quality. Low doses in the range 50–370 Gy were found by Alabastro et al. (1978) to significantly reduce *Colletotrichum* decay in Phillipine Cavendish bananas, probably due to ripening inhibition at this dosage. No effect, however, was observed on *Thielaviopsis*, Applying doses of 0.3–1.5 kGy to South African bananas, Huyzers and Basson (1985) found that shelf-life extension of 2–3 weeks could be obtained at the higher doses. Although no significant changes in sensory quality were found even at the higher doses, signs of radiation damage were already evident at a dose of 0.7 kGy. A large-scale marketing trial with mature, green bananas treated shortly after harvesting with 0.4 or 0.5 kGy indicated that losses had decreased from 18 to 2% and sales had increased by 20% (Huyzer and Basoon 1985). In large-scale trials with South African bananas, Brodrick and Strydom (1984) and Brodrick et al. (1984) achieved a boubling of storage life in the ripening rooms (from 14 to 29 days) with an average dose of 0.85 kGy. At this dose, however, a slight phytotoxic effect to the fruit was noticed. Radurization at 0.60 kGy resulted in an extension of 2–4 days in the marketable life, and browning associated with handling of ripe fruit was also reduced. As a result of these experiments one of the major chain stores in Johannesburg has been regularly marketing irradiated fruit.

176   Case Studies

Other Fruits. Fresh figs are highly perishable as they are subject to attack by surface molds. Bramlage and Couey (1965) found that radiation at 2–4 kGy reduced the incidence of decay in three fig cultivars by retarding disease development rather than preventing it. Decay reduction occurred both under cold storage and shelf conditions. However, the longer the fruits were held in cold storage (3.3 °C) the shorter was the radiation-induced extension of shelf-life. Radiation damage of figs was found to be cultivar dependent as shown by Kadota figs from different growing areas but not by Calimyrna or Mission figs. No effect of radiation on flavor of any of the figs was recorded.

For raspberries (cv. Heritage) harvested at the pink or red stages of ripeness, irradiation at 1–3 kGy substantially reduced yeast populations, but the pathogenic fungi *Penicillium* and *Cladosporium* were less sensitive (Larrigaudière et al. 1987). At 1 kGy no reduction in firmness was recorded and storage life of both ripe and unripe fruit at 4 °C was extended.

Working with six cultivars of lychees, Moy et al. (1969) found that irradiation at 0.5 and 1 kGy decreased the incidence of decay without affecting the aroma, flavor, texture, and color of the pulp. Akamine and Goo (1977b) found that the tolerable level of irradiation by lychees is limited by pericarp darkening during cold storage (1.7 °C) and by decay above 7.2 °C. The optimum combination of radiation dose and storage temperature was found to be 250 Gy and 7.2 °C for shelf-life maintenance but not for shelf-life extension.

Irradiation of Egyptian dates at 1 and 2 kGy greatly reduced postharvest decay caused by *B. cinerea* and *P. expansum*, and extended storage life at 22–30 °C by 8 days, without affecting percent of soft fruit and levels of sugar, protein, and total amino acids (El-Sayed 1978b).

Tomatoes. The development of fungal decay, caused mainly by *A. alternata*, *R. stolonifer*, and *B. cinerea* (Barkai-Golan 1981; Dennis 1983), and bacterial decay, caused by *Pseudomonas* and *Erwinia* species (Bartz 1980; Spalding and Reeder 1986a), are the major factors limiting the extension of storage and marketing of tomatoes. Although most irradiation studies with tomatoes were aimed at extending postharvest life by delaying ripening, some reports have shown an increase in shelf-life due to decay reduction. Bramlage and Lipton (1965) reported that storage decay of some tomatoes was reduced by 3 kGy, but the treatment caused mottling (an uneven red pigmentation) and softening of the fruit. Doses of 1–3 kGy were also found by Salunkhe (1961) to prevent even ripening and to impair red color development in green tomatoes. Mathur (1968) considered a dose of 750 Gy optimum for decay control and avoidance of tomato injury in tomatoes stored at 9–10 °C and 85–90% relative humidity. A dose of 1 kGy adversely affected ripening although decay was controlled. Abdel-Kader et al. (1968) found that shelf-life of two tomato cultivars (Early Pak No. 7 and Ace)

Radiation Effects on Disease Development 177

irradiated at 2.5–3 kGy and stored at 12–15 °C was extended by 4 to 12 days due to decay control, provided that the fruits were at their pink or riper stage when irradiated. Doses higher than 3 kGy caused physiological breakdown, resulting in more decay, while lower doses were ineffective for decay control. Irradiation at 250–1000 Gy was found by Spalding and Reeder (1986a) to reduce the incidence of bacterial soft rot in mature, green tomatoes inoculated with *Erwinia carotovora*, *Erwinia* sp. and *Pseudomonas fluorescence*. Reduction in *Erwinia* soft rot was greatest when tomatoes were irradiated at 1 kGy within 3 h after inoculation. Irradiation at 0.5 or 1 kGy, however, increased the incidence of fungal decay, although no peel damage was observed.

Potatoes. Irradiation at 70–100 Gy is an effective method of inhibiting sprouting in potato tubers (Thomas 1983). However, it was generally found that irradiation at sprout-inhibitory doses may result in enhanced susceptibility to storage pathogens. Attempts to suppress decay by higher radiation doses showed that less than 0.5 kGy was sufficient to prevent decay caused by the radiosensitive *Phytophthora infestans*, without causing any damage to the tubers (Beraha et al. 1959c; Beraha et al. 1960). However, *Pythium debaryanum* and *Corynebacterium sepedonicum* were not completely inactivated by 1.5–2 kGy, and doses of 4.5–5 kGy failed to prevent decay by *E. carotovora* and *Fusarium* spp. (Beraha et al. 1959c; Duncan et al. 1959; Hooker and Duncan 1959; Beraha et al. 1960).

Onions and Garlic. Irradiation at sprout-inhibiting doses (20–120 Gy) has been reported to intensify storage rots in onion and garlic (Dallyn and Sawyer 1954; Nuttall et al. 1961; Skou 1971). However, in contrast to potatoes, reports on decreased or unchanged incidence of rot have also been recorded (van Kooy and Langerak 1961; Mullins and Burr 1961; Mumtaz et al. 1970). The reason cannot be the direct effect on the pathogen, since the main decay microorganisms, such as *Botrytis allii*, *B. cinerea*, *Aspergillus* spp., *Penicillium* spp., and others, are almost unaffected by these low doses (Beraha et al. 1960; Saravacos et al. 1962; Beraha 1964; Sommer 1964a; Barkai-Golan et al. 1967). Several studies led to the conclusion that when well-cured and healthy bulbs of good quality are irradiated during the dormancy period with the minimum doses needed for sprout inhibition, and when good storage management is practiced, rot is not significantly increased (Matsuyama and Umeda 1983; Thomas 1984).

Carrots. Irradiating carrots at a sprout-inhibition dose (120 Gy) generally resulted in increased decay during 6 months' storage at 2 °C and 95–100% relative humidity, although less decay was recorded in washed than in unwashed carrots (Skou and Henriksen 1964). For short-term storage of less than a month, however, irradiation seemed to be of advantage in decay

178   Case Studies

suppression (Skou 1977). In contrast to gamma rays, fast electrons (β rays) applied to the top ends only, prevented sprouting for 6 months, while the incidence of rot did not differ from that of the nonirradiated product (Skou 1977).

Mushrooms. Cultured mushrooms (*Agaricus* spp.) are highly perishable and can be kept in prime condition for only 1 day at 10°C or 5 days at 0°C. Extensive studies indicated that radiation can extend the market life of *Agaricus* mushrooms by suppressing cap opening, stalk elongation, darkening of the gills, cap, and stalk, shrivelling, and surface mold development, without impairment of quality (Bramlage and Lipton 1965; Campbell et al. 1968; Wahid and Kovacs 1980; Roy and Bahl 1984a,b; Smierzchalska and Wojniakiewicz 1986; Beelman 1988). An irradiation dose as low as 63 Gy was sufficient to inhibit cap opening but only doses of 500 Gy were effective in maintaining the fresh appearance of mushrooms and extending their marketable period (Bramlage and Lipton 1965; Barkai-Golan and Padova 1975). At this dose respiration rate under shelf conditions (15°C) was reduced by 50%. A further increase in the radiation dose did not reduce the rate of respiration any more (Barkai-Golan and Padova 1975). Doses in the range 0.25–4 kGy were effective in preventing mold growth during storage (Bramlage and Lipton 1965). Roy and Bahl (1984a) found that irradiating freshly harvested mushrooms, held in perforated polyethylene bags, at 2.5– 5.5 kGy considerably reduced cap opening and prevented mycelial growth and spoilage during storage at 15°C. Both gamma irradiation at 2 kGy and high-energy electrons at 1 and 2 kGy retained the good quality of Polish *Agaricus* mushrooms for 6–8 days at 10–18°C (Smierzchalska and Wojniakiewicz 1986; Smierzchalska et al. 1988, 1989). Along with delay in postharvest growth, gamma irradiation also resulted in reduced infection by *Acremonium album* and *Pseudomonas tolaasi*. Inconsistent effects of radiation on color and quality of Chinese *Pleurotus* mushrooms (*Pleurotus ostreatus*) wrapped in polyethylene bags were reported by Lu (1988) at doses up to 800 Gy. However, quality maintenance was always better at low temperatures and was cultivar dependent.

Other Vegetables. Irradiation with X-rays at 1–3 kGy was sufficient to inhibit growth of *Fusarium*, *Cephalosporium*, *Alternaria*, and *Stemphylium* species on sweet red peppers, whereas a dose of 4–8 kGy was necessary for the control of these organisms (Bramlage and Lipton 1965). However, this amount of irradiation stimulated softening, yellowing, and calyx discoloration and its application for decay control is not recommended.

Irradiation at 3 kGy reduced decay development on stem scars of some cucumbers but not on others; however, all fruits were softened and the fresh green appearance was impaired (Bramlage and Lipton 1965). Loss of chlorophyll and firmness have already been reported by Morris et al. (1964)

Radiation Effects on Disease Development

at doses above 500 Gy. Bramelage and Lipton (1965) found that decay of summer squash and cantaloups was not consistently affected by radiation, whereas softening of the vegetables was increased. Ravetto et al. (1967) indicated that irradiation at doses tolerated by cantaloups was not beneficial for the preservation of the commodity, whereas doses above 4 kGy predisposed it to attack by storage pathogens.

Radiation at 0.5–2 kGy reduced decay of some lettuce cultivars but not of endive. Induced injury, however, was so great as to preclude irradiation as a postharvest treatment (Bramlage and Lipton 1965). Radiation at 1– 4 kGy was also inefficient for reducing *Botrytis* rot of globe artichokes and resulted in stem pitting and discoloration which increased with dose (Bramlage and Lipton 1965).

Sweet potatoes irradiated at 2 or 3 kGy were severely injured and after 3 months' storage all the roots were decayed (Bramlage and Lipton 1965). Following irradiation at sprout-inhibition doses it was found that 82.5 and 165 Gy enhanced decay only when the roots were stored at chilling temperatures, while 250 Gy increased decay regardless of the storage temperature.

Combined Treatments

For most fruits and vegetables even doses which are sublethal for the pathogen, but are capable of retarding its growth or halting it temporarily, may result in radiation-induced damage to the host. In an attempt to reduce the radiation doses required for disease control beneath the threshold of damage, combined effects of radiation with other physical or chemical treatments have been investigated (Barkai-Golan et al. 1977).

Heat and Radiation Treatments. Heat and gamma radiation was found to react synergistically on spore inactivation of various postharvest fungi (Sommer et al. 1967, 1968; Barkai-Golan et al. 1969a; Ben Arie and Barkai-Golan 1969; Brodrick and Thomas 1978; Barkai-Golan and Padova 1981) leading to a decreased required intensity for each of the treatments when applied separately. The magnitude of the effect for a given pathogen is influenced by the sequence of treatments, a greater effect being generally observed when heat preceded irradiation, suggesting that heating sensitizes fungi to radiation. Thus, with the brown rot fungus, *Molilinia fructicola*, Sommer et al. (1967) found a five- to tenfold synergism above the additive effects of the two treatments when heat preceded irradiation. However, studies by Buckley et al. (1967) with nongerminating and germinating sporangiospores of *R. stolonifer* indicated that maximum synergism for the resistant nongerminating spores was obtained with a radiation–heat sequence, while for the more sensitive germinating spores maximum synergism was achieved by a heat–radiation sequence.

180   Case Studies

For some fungi the two treatments are complementary. Thus, for *P. expansum*, which is relatively heat resistant but radiation sensitive (Sommer et al. 1967; Ben Arie and Barkai Golan 1969), or for *A. alternata*, which is heat sensitive but radiation resistant (Ben Arie and Barkai-Golan 1969), a mild heat treatment followed by a low radiation dose considerably suppressed, or even completely inhibited, spore germination. Heating followed by irradiation reacted synergistically in suppressing germination of *Aspergileus ochraceus* sclerotia (Paster and Barkai Golan 1986). The combined treatment, however, did not consistently exhibit a synergistic effect on ochratoxin production by the sclerotia, suggesting that the heat–radiation combination may have a different effect on physiological processes (germination) than on synthetic pathways (ochratoxin production) in the same fungus. The synergistic effect on mycelial growth was demonstrated by Brodrick and Thomas (1978) with the fungus *Hendersonia creberrima*, the causal organism of soft rot in South African mangos. In this case, heating at 55 °C for 2.5, 5, or 10 min, followed by irradiation at 0.75 kGy, resulted in complete inhibition of mycelial growth. Another factor which may affect the rate of synergism is the time interval between heating and radiation. It was generally stated that in order to obtain effective fungal control, irradiation should be applied within 24 h of the hot water treatment (Brodrick and van der Linde 1981; Spalding and Reeder 1986b).

Recovery studies by Matsuyama (1978) with *Escherichia coli* cells of different DNA repair capacities, suggested that the synergistic interaction of heat and irradiation may involve the inhibition of DNA repair and the recovery of cells from the heat damage.

The suppressive effect of a heat–radiation treatment in vivo will depend not only upon the sensitivity of the pathogen to heat and radiation and to the synergistic effect obtained but also upon the extent of infection. As with single radiation or heat treatments, the effect of combined treatment in vivo will be limited by host tolerance and should be evaluated for each host–pathogen interaction.

Strawberries. Heating strawberries at 41 °C with humidified air, prior to irradiation at 2 kGy, prevents contact infections from diseased to healthy berries for 10 days at 5 °C (Sommer et al. 1968). However, the heat treatment designed for sensitizing the fungus must be carefully controlled, and the time of exposure to elevated temperatures without adverse physiological effects on the fruit is short. *Rhizopus stolonifer*, the pathogen of leak disease, is more resistant than *B. cinerea* to both heat and radiation (Sommer et al. 1967). This pathogen, however, does not develop under refrigeration and it was concluded that heat and radiation treatments should not be expected to overcome problems of inadequate refrigeration (Sommer et al. 1968).

Radiation Effects on Disease Development 181

Rhizopus stolonifer and, to a lesser extent, *Colletotrichum acutatum* are the main problems affecting locally marketed strawberries in South Africa (Brodrick et al. 1977). Working with Parfait, Selekta and Tioga cultivars, Brodrick et al. (1977) found that moist heat at 50–52 °C for 10 min plus irradiation at 2 kGy effectively controlled both pathogens without adversely affecting berry quality. In semicommercial experiments the combined treatment effectively controlled fungal diseases for several days after picking, thus allowing sufficient time to market the fruit under local market conditions.

Stone Fruits. Working with *M. fructicola*-inoculated nectarines (cultivars Late LeGrand and Gold King) Sommer et al. (1967) found that irradiation at 1 kGy or a hot water dip at 55 °C for 3.5 min were ineffective in reducing the incidence of brown rot when applied separately, whereas the combination of the two treatments entirely prevented decay for 5 days at 5 °C and for 10 days at 20 °C. Synergistic effects were also observed in Fiesta peaches when heating at 50 °C for 3.5 min followed by a 1 kGy dose was sufficient to inhibit lesion development for 5 days at 20 °C. The advantage of heattreatment before irradiation was demonstrated in peaches by Dennison and Ahmed (1971). After 10 days at 2 °C plus 4 days at 20 °C, irradiation at 1.5 kGy resulted in 2.2% decay, and heating at 49 °C for 7 min resulted in 4.4% decay. However, no decay was recorded in stored fruit receiving the combined treatment.

The effect of a combined treatment was studied by Brodrick et al. (1985) on South African cultivars of plums and nectarines inoculated with *R. stolonifer*, *M. fructicola*, and *B. cinerea*. Although the pathogens could be suppressed by severe hot water treatment (46 °C for 10 min), unacceptable fruit damage occurred, as expressed by a general shrivelling and softening as well as by cracking at the distal end. Similarly, irradiation alone at 2 kGy could control fungal development but resulted in slight softening of the fruit. Mild heat treatment of 42 °C for 10 min followed by irradiation at 0.75–1.5 kGy effectively controlled fungal development with no significant changes in fruit texture, aroma, or taste.

Pome Fruits. Working with three cultivars of apple (Cox, Lacton, and Lambarts) Langerak (1982) found that a hot water dip at 45 °C followed by irradiation at 1.25 kGy considerably reduced the percent of decay in *P. expansum*- and *B. cinerea*-inoculated fruit stored for 5 months (as the postclimateric stage) at 4–5 °C. Irradiation at 0.5 kGy applied with heat treatment was sufficient to prevent decay in fruits treated at the beginning of the storage period (at the preclimacteric stage). The combined treatments, however, had a slight adverse effect on the sensory properties of the apples.

182   Case Studies

Decay in Spadona pears inoculated with *P. expansum* was almost completely prevented by a hot water dip at 47°C for 7 min followed by irradiation at 0.5 kGy. The same treatment did not prevent rotting of fruits inoculated with *B. cinerea* and *A. tenuis* (*A. alternata*), although a delay in disease initiation was observed (Ben Arie and Barkai-Golan 1969).

Grapes. Padwal-Desai et al. (1973) showed that the combination of a hot water dip (50°C for 5 min) and gamma irradiation at 0.1 kGy considerably extended the shelf-life of Seedless and Anab-e-Shai grapes packaged in perforated polyethylene bags, by reducing *Rhizopus* and *Aspergillus* decay. The treated berries retained their texture and flavor, and the organoleptic rating was comparable to that of fresh untreated samples. Combined heat–radiation treatment has also resulted in good control of *B. cinerea* in South African table grapes (Brodrick 1982). However, moisture remaining within the bunch following a hot water dip frequently resulted in increased *Penicillium* infections. Dipping the fruit in a 30% ethanol solution after hot water treatment to facilitate drying successfully controlled decay by both *Botrytis* and *Penicillium*.

Citrus Fruits. A significant reduction in the incidence of decay was observed by Ahmed et al. (1968) in several citrus varieties (Pineapple, Temple, and Valencia) by combining a hot water dip (52.7 or 56°C for 5 min) and gamma irradiation at 1.1–2.1 kGy. These high doses, however, resulted in peel injury. Working with *P. digitatum*-inoculated Shamouti oranges, Barkai-Golan et al. (1969a) found that 0.5 kGy applied 1.5 h after immersion in water at 52°C for 5 min delayed the appearance of rot by 33–40 days. Only 2.5–5% of the fruits showed rot during the 50-day holding period at 14°C, whereas all the untreated fruits rotted after 6 days. Fruits receiving irradiation alone rotted within 16 days, whereas most of the heated fruits rotted within 34 days. Heating at 50°C plus irradiation at 350 or 500 Gy reacted synergistically on suppressing blue mold development in *P. italicum*-inoculated grapefruits. Reducing the radiation dose in the combined treatment had less effect on shortening the incubation period of the disease than did lowering the temperature. The combined treatments were less effective on Shamouti oranges inoculated with *P. italicum*, suggesting that for each host–pathogen system there is an optimal combination treatment. Considerable reduction in decay development due to synergistic effects was also exhibited by applying heat treatment (52°C for 5 min) plus electron irradiation (500 Gy of 520 keV electrons) (Barkai-Golan and Padova 1981) to Valencia oranges superficially inoculated with *P. digitatum* spores. With a longer time lag between inoculation and irradiation the effects of the combined treatments were less pronounced.

In spite of the reduction in radiation dose due to the combined treatments (Barkai-Golan 1969a, 1973), peel damage has sometimes been ob-

Radiation Effects on Disease Development   183

served in oranges and grapefruits treated with doses as low as 250–500 Gy (Barkai-Golan and Padova 1970b, 1981). Hatton et al. (1982, 1984) found that doses of 600 and 900 Gy often resulted in tissue breakdown and scald in Florida grapefruits stored at 10–16 °C. Only at doses of 300 Gy or lower was no injury, or at least minimal injury which resulted in an acceptable fruit, recorded. Differences in appearance, flavor, odor, and taste were observed by O'Mahony and Goldstein (1987) in Navel oranges irradiated at 0.52–0.60 kGy, whereas differences in fruits subjected to 0.32–0.37 kGy were less extreme. However, several studies emphasized that a hot water dip, polyethylene wrappers, or waxing prior to irradiation decreased injury symptoms and resulted in improved fruit appearance (Ahmed et al. 1968; Kahan and Padova 1968; Barkai-Golan and Padova 1981).

Avocado. Ripening delay in avocados due to combinations of low radiation doses (10–30 Gy) with mild hot water treatments (45–46 °C for 10 min) extended shelf-life of these radiation-sensitive fruits by 2–5 days under commercial marketing conditions (Thomas 1977). Chilean Fuerte avocados treated by a hot water dip (45 °C for 5 min) prior to irradiation (25 Gy) and wrapped in PVC shrink foil were firmer with a better appearance and flavor than nonirradiated fruit, after 4 weeks transport at 7 °C from Chile to Europe. Furthermore, this treatment reduced stem-end rot when the fruit was stored at 12 °C on arrival at its destination (Langerak 1984).

Mango. Irradiation of mangos at dosages of 1.1–2.1 kGy reduced the incidence of anthracnose (*C. gloeosporioides*) to some extent, but improved control was obtained with irradiation combined with hot water treatment (Ahmed et al. 1968; Dennison and Ahmed 1971). Very satisfactory disease control was reported by Brodrick and Thomas (1978) in mango shipment trials from South Africa to Europe by combining irradiation at 0.75 kGy with hot water (55 °C for 5 min) or hot benomyl treatment. Such a combination has been used commercially for treating mangos in South Africa and was reported (Thomas 1975) to act synergistically in controlling anthracnose caused by *C. gloeosporioides* and soft brown rot caused by *Hendersonia creberrima*, in addiation to providing quarantine control of the mango seed weevil (*Sternochetus mangiferae*). Spalding and Reeder (1986b) found that the combination of a hot water dip (53 °C for 3 min) with radiation at 0.2 or 0.75 kGy were more effective than single treatments for control of anthracnose (*C. gloeosporioides*) and stem-end rot (*D. nathalensis* and *P. citri*) in Tommy Atkins mangos. The most effective decay control was achieved with hot water plus 0.75 kGy when 88% of the fruit were acceptable after 17 days at 13 °C followed by ripening at 24 °C. This dose, when applied alone, induced skin injury and inhibited development of ripe skin color in various mango cultivars (Hatton et al. 1961; Farooqui et al. 1974a; Spalding 1986b). However, when irradiation was preceded by heat treat-

184 Case Studies

ment the injurious effect was considerably reduced and the inhibitory effects on skin color were partially offset (Spalding 1986b).

Papaya. Promising results from the use of combined heat–radiation treatments have been reported for Solo papayas by Moy et al. (1973). The shelf-life of the fruit, which was affected both by the rate of fruit ripening and decay development, was extended by 3–31/2 days by immersion in hot water (49 °C for 20 min) prior to irradiation at 0.75 kGy. It was suggested that the irradiation delayed ripening of the fruit by retarding respiration, while the heat treatment directly controlled storage decay.

Hunter et al. (1969) indicated that irradiation alone has a limited fungistatic effect on stem-end rot (mainly *Ascochyta caricae*) at a level noninjurious to the fruit (0.75–1 kGy), whereas a hot water dip (48–50 °C for 20 min) prior to irradiation effectively controlled decay. However, heat–radiation treatment increased scalding injury above that caused by heating alone. The combination of irradiation at 0.75 kGy with conventional hot water treatment (10 min at 50 °C) was found to facilitate the distribution of South African Papino papayas within the country and made large-scale export by sea possible (Brodrick et al. 1976; Brodrick and Thomas 1978). An average shelf-life extension of 9 days longer than that for heated but nonirradiated control fruits was achieved under simulated export conditions. In this experiment most of the control fruit was rejected due to disease development, mainly anthracnose (*C. gloeosporioides*) and stem-end rot (*Phoma* sp.) and, to a lesser extent, *R. stolonifer* rot. The decay during shipment could be controlled by the heat treatment alone, but nonirradiated fruits overripened and collapsed rapidly when kept at 20 °C. Disease control was less effective when the radiation dose was increased to 0.1 kGy. Thus, although no differences in fruit firmness or flavor could be detected at 0.1 kGy, this treatment was considered to be slightly phytotoxic to the fruit.

Banana. The advantage of a heat–radiation treatment for postharvest disease control in bananas has been demonstrated by Padwal-Desai et al. (1973). Immersion of green preclimateric Mysore and Dwarf Cavendish bananas in hot water (50 °C for 5 min) plus irradiation at 0.25 and 0.35 kGy respectively, reduced the incidence of stem-end rot caused by *Gloeosporium musarum* and, at the same time, delayed ripening during storage at 28–32 °C or at 15 °C. Disease control was more pronounced when the hot water dip preceded irradiation than vice versa.

Tomatoes. A hot water dip at 50 °C for 2 min followed by irradiation at 0.5 kGy was found by Barkai-Golan (unpublished data) to totally eliminate *A. alternata* decay in light red tomatoes under natural infection conditions at 23 °C. Single heat or radiation treatments resulted in a 90 or 40%, respectively, reduction in decay. In *B. cinerea*- and *R. stolonifer*-inoculated tomatoes decay reduction

Radiation Effects on Disease Development

was achieved due to the synergistic action of heating and irradiation at 1 kGy. Reduced incidence of decay by *P. expansum* and *R. stolonifer* in tomatoes was reported by El Sayed (1978a) with irradiation at 1 kGy preceded by a hot water dip at 60 °C for 2 min. Such a treatment extended shelf-life by 13 days without affecting vitamin C, sugar, or amino acid contents. Spalding and Reeder (1986a) found that gamma irradiation at 0.5–1 kGy, but not hot water treatment (57 °C for 2 min), reduced the incidence of bacterial soft rot caused by *Erwinia* and *Pseudomonas*. The combination of 0.75 kGy with heating, before or after irradiation, provided no better control of bacterial rot than irradiation alone. Thus, the synergistic effect of the combination reported for control of fungal decay of tomatoes (El-Sayed 1978a; Barkai-Golan et al. 1991) was not operative for bacterial soft rot pathogens.

Chemical and Radiation Treatments. Studying the possibility of increasing fungal radiosensitivity by chemical treatments, Georgopoulos et al. (1966) found that iodoacetamide, in the presence of oxygen, was a very effective radiosensitizer for *P. italicum*, *R. nigricans* (*R. stolonifer*), *B. cinerea*, and *A. pullulans*. In combination with this chemical, radiation doses as low as 40 Gy destroyed not only the colony-forming potential of *B. cinerea* but also the germinability of the spores, which are much more resistant to radiation than the former (Sommer et al. 1963a). Sensitizing effects were also found for iodoacetic acid when tested on *P. italicum* and *A. pullulans* spores. On the other hand, some of the compounds tested failed to sensitize the microorganisms, while others exhibited considerable toxicity (Georgopoulos et al. 1966). Cooper and Salunkhe (1963) found that preirradiation treatment of Shasta strawberries with captan (1000 ppm) or potassium sorbate (2000 ppm) enhanced the effect of radiation and increased the percent of marketable fruit stored at 4.4 °C. Captan had a better antifungal effect in conjunction with radiation and resulted in 65 and 100% marketable fruit at 2 or 3 kGy, respectively. At the higher dose, however, the strawberries turned spongy and became water-soaked. For Bing cherries, combinations of chemical dips, of captan (1000 ppm), mycostatin (100 ppm) or myprozine (100 ppm), with irradiation (3 or 4 kGy) were more effective than the chemical treatment alone in increasing the percent of marketable fruit during 80 days' storage at 4.4 °C.

Roy and Mukewar (1973) found that whereas irradiation alone at 2 kGy had very little fungicidal effect on black mold, *A. niger*, of apple the coupling with aureofungin or captan considerably retarded or totally inhibited colony growth, respectively. Irradiation at 2 kGy of *A. niger*-inoculated apples followed by a dip for 10 min in 1000 ppm aureofungin or captan markedly prolonged the incubation period of the disease and protected 85% of the fruit for 3 weeks at 20 °C. The combined treatment of radiation (2 kGy) with benomyl (500 ppm) was similarly more effective than the single treatments in suppressing *Fusarium coeruleum*, the potato dry rot fungus (Roy and Mukewar 1973). Combination of captan treatment (20 mg/

186 Case Studies

l) with irradiation at 2 or 3 kGy was found by Georgiev (1983) to reduce the disease index in *B. cinerea*-infected Bolgar grapes from 99.00 to 8.67 after 60 days' storage at 4–10 °C. These high radiation doses, however, affected fruit color and turgor. Applying 1 kGy in the combined treatment reduced the index values from 99.00 to 12.74 without affecting fruit quality.

The effect of combined action of gamma radiation with the fungistatic compound diphenyl or the fungicidal compound sodium orthophenylphenate (SOPP) was studied on the in vitro growth of citrus pathogens (Barkai-Golan and Kahan 1967; Kahan and Barkai-Golan 1968). Barkai-Golan and Kahan (1967) found that irradiation at 0.6 kGy and treatment with only 5 mg diphenyl halted mycelial growth for 14 days at 25 °C of the diphenyl-sensitive *P. digitatum* and *D. natalensis*, and of *T. viride*, which is diphenyl-resistant but very radiosensitive. A combination of 5 mg diphenyl with a dose of 1.4 kGy was required to achieve a similar effect for *P. italicum*.

For *P. citrophthora* and *Oospora citri-auraunti* (*Geotrichum candidum*), which are diphenyl-resistant but radiation-sensitive, the combined treatments did not further improve the effect of radiation alone. Similarly, the combined action did not increase the suppressive effect of diphenyl on *C. gloeosporioides* and *A. citri*, which are radiation-resistant but react to some extent to diphenyl.

Kahan and Barkai-Golan (1968) showed that sublethal doses of 0.6–1.4 kGy considerably increased the in vitro sensitivity to SOPP of all the citrus pathogens tested, as was exhibited by the reduced SOPP concentration required for colony growth inhibition. Incorporation of SOPP at 0.0025% in wax coatings plus gamma radiation at 1 kGy was reported by Ahmed (1977) to extend the shelf-life of Balady oranges by 15 weeks at 14–20 °C, as compared to 7–10 weeks in the untreated or nonirradiated, coated fruit.

Incidence of stem-end rot (mainly *Ascochita caricae*) in Solo papayas was reduced by about 30% when 2-aminobutane treatment was combined with irradiation at 0.75–1 kGy, as compared to the effect to the chemical alone. However, the addition of irradiation resulted sometimes in increased scalding injury (Hunter et al. 1969).

Combinations of low radiation doses (50 Gy) with fungicidal dips were found by Wu et al. (1980) to be more effective than fungicide treatment alone in reducing diseases (*Aspergillus niger*, *Fusarium* spp. and *Erwinia* spp.) in onion bulbs during storage for 30–90 days at 10 °C.

Heat, Chemical and Radiation Treatments. Several studies have shown the advantage of the combination of low radiation dose, mild heat treatment and chemicals over the double-component combined treatments in controlling decay during storage. Roy (1975) found that the association of heating at 50 °C for 10 min, 1.5 kGy irradiation and treatment with 250 ppm benomyl, and heating at 56 °C for 4 min, irradiation with 1.5 kGy and

Radiation Effects on Disease Development

treatment with 1000 ppm aureofungin, in these sequences, completely controlled 2-day-old infections of *P. expansum* in apples during the 3-week holding period at 25 °C. For *P. digitatum*-inoculated Shamouti oranges, Barkai-Golan and Padova (1970a) showed that the combination of irradiation (200 Gy), diphenyl (15 mg per fruit) and a hot water dip (52 °C for 5 min) extended the incubation period of green mold to 14 days at 14 °C. An incubation period of 8 days was recorded in fruits treated with heat plus radiation or with diphenyl plus radiation, while a 6-day incubation period was found for heat-treated, irradiation-treated or nontreated fruit. Nyambati and Langerak (1984) found that the incubation period of green mold in *P. digitatum*-inoculated Persian limes, stored at 15 °C, was extended to 12 days by a hot water dip (45 °C for 5 min) plus treatment with $K_2S_2O_5$ (0.5%) and irradiation at 250 Gy, as compared to 8 or 6 days in the heat- and chemical-treated fruits respectively, and to 4 days in fruit irradiated at 500 Gy. Working with Tommy Atkins mangos, Spalding and Reeder (1986b) found that the most effective forms of decay control were achieved by the combination of hot water (53 °C for 3 min) and radiation (0.2 or 0.75 kGy) or by hot imazalil (0.1 active ingredient) and radiation. The addition of imazalil to a heat−radiation treatment did not further decrease fungal decay while its absence had the advantage of leaving no chemical residues with the fruit.

Jacobs et al. (1973) found that both anthracnose (*C. gloeosporioides*) and soft brown rot (*H. creberrima*) were successfully controlled in mangos by submerging the fruit immediately after picking in benomyl suspension at 55 °C for 5 min. The loss of fruit luster after heating could be overcome by waxing the fruit. The addition of irradiation at 1 kGy after a hot benomyl dip plus wax application did not further improve the control obtained by hot fungicide alone.

Irradiation doses ranging from 0.3−1.2 kGy were found by Johnson et al. (1990) to reduce postharvest disease in Kensington Pride mangos but the level of control was not commercially acceptable. Hot benomyl treatment prior to irradiation resulted in an additive effect and provided effective control of anthracnose (*C. gloeosporioides*) and stem-end rot (*Dothiorella dominicana*) for 15 days at 20 °C.

Irradiation and Controlled Atmosphere. Various combinations of CO_2 (up to 10%) and O_2 (more than 5%) were found by Chalutz et al. (1965) to be less effective than irradiation at 2 kGy in suppressing *B. cinerea* in strawberries. The advantage in combining a controlled atmosphere with irradiation was considered to be insufficient to justify the additional expenditure. Looking for methods to reduce decay in Kensington Pride mangos, Johnson et al. (1990) found that hot benomyl dip (52 °C for 5 min) plus irradiation at 0.3−1.2 kGy effectively controlled decay during short-term storage. Satis-

188   Case Studies

factory disease control was achieved during long-term controlled-atmosphere (5% O_2, 1.5–2% CO_2) storage when the fruit was treated with hot benomyl followed by prochloraz and irradiation.

Irradiation and Polymeric Film Packaging. Packaging of peaches (Alberta and Gem cultivars) in different polyethylene bags resulted in increased CO_2 concentrations inside the bags, the level of which differed with the type of film. Dhaliwal and Salunkhe (1963) found that fast electron radiation at $1–5 \times 10^5$ rad effectively suppressed the in vitro growth of peach pathogens (*Penicillium*, *Rhizopus*, and *Alternaria* species). Fruits subjected to this dosage were more sensitive to fungal attack and were infected earlier than those subjected to lower doses. Combination of fast electron or gamma irradiation at $1–3 \times 10^5$ rad with packaging in polyethylene bags increased the effect of radiation alone and extended the refrigerated life of the fruit by 15–20 days. Such a packaging was responsible for the accumulation of an increased level of CO_2 within the bags.

Looking for methods to broaden the marketability of citrus fruits which suffer from peel damage (e.g., mechanical injury, oleocellosis, or various physiological disorders), Barkai-Golan et al. (1984) studied the effect of irradiation on peeled citrus fruits. Gamma radiation at 1.2 kGy reduced by 50% the incidence of rot on peeled Shamouti oranges inoculated with *P. digitatum* and stored at 17 °C. Combining irradiation with packaging in sealed high density polyethylene or polyvinyl chloride wrappers resulted in 25 and 10% rot respectively, and this treatment reacted additively in decreasing contamination by *Cladosporium herbarum* of the peeled fruit. Both high density polyethylene and polyvinyl chloride films delayed dryness of the fruit under shelf-life conditions. On the other hand, Heins and Langerak (1980) noted that suboptimal packaging conditions may reduce the effect of irradiation alone or of combined heat–radiation treatment; this was observed for papayas and strawberries in an airborne trial shipment from South Africa.

Ultraviolet and Gamma Radiation. The combination of ultraviolet with gamma radiation to impede microbial growth was studied by Moy et al. (1978) on *Phytophtora*, *Colletotrichum* and *Ascochyta* species, the prevalent pathogens of papayas in Hawaii. It was suggested that ultraviolet radiation followed by gamma radiation would cause breaks in the double strand of microbial DNA, which could not be repaired by photoreactivation or other comparable means (Davis et al. 1970). Species of *Phytophtora* were more radiosensitive than the other fungi and gamma radiation at 0.25 kGy plus ultraviolet radiation at 3.9×10^4 erg/mm^2 was sufficient to prevent growth. A combination of 1.5 kGy gamma radiation plus 1.19×10^4 erg/mm^2 ultraviolet radiation was required to control *Colletotrichum* spp., which survived

Pathological and Microbiological Problems Following Irradiation 189

gamma radiation at doses up to 2.5 kGy. For *Ascochyta*, the most resistant genus of the three fungi, gamma radiation at up to 2.5 kGy and ultraviolet radiation at up to 7.33×10^4 erg/mm^2, alone, or in combination, was not sufficient to completely prevent growth and colony formation.

5.1.4 Pathological and Microbiological Problems Following Irradiation

Increased Susceptibility to Infection. Irradiation can be beneficial for the preservation of fruits and vegetables only in the limited dose range which is sufficient for decay suppression but which is below the threshold of damage to the host. Unsuitable doses may result in chemical, physiological, or textural damage as exhibited by peel injuries, anomalous ripening, softening of the tissues, or changes in aroma and flavor. Following these changes, injury-inducing doses frequently lead to an enhanced decay development due to the reduced resistance of the tissues to invasion by the weak postharvest pathogens. Softening of the tissues, which is associated with the degradation of cell wall pectic constituents and increased permeability of host tissues (Skou 1964b; Massey et al. 1965), may be responsible for the enhanced invasion of the fruit by pectolytic pathogens. Furthermore, by softening the tissue, irradiation may indirectly expose the fruit to wounding and thus to infection by wound pathogens.

For mango fruits, Alabastro et al. (1978) found that doses of 160 and 190 Gy were more effective in delaying the appearance of anthracnose and stem-end rot than a 220 Gy dose. This was suggested to be the result of the reduction in fruit resistance to infection due to induced damage at the higher dose. For various grape cultivars a reduction in fruit resistance to repeated infection was recorded only at doses above 4 kGy, whereas 2–3 kGy doses resulted in storage-life extension (Rogachey 1966). Studies by Ravetto et al. (1967) on cantaloups showed that doses tolerated by the fruit were not beneficial for preservation of the commodity, whereas doses above 4 kGy predisposed it to storage pathogens.

In Shamouti oranges inoculated with *P. digitatum* a hot water dip (52 °C for 5 min) plus irradiation at 0.5 kGy considerably suppressed decay development during 50 days of storage. Increasing the dose to 1 kGy led, however, to increased decay incidence (Barkai-Golan et al. 1969a). A similar phenomenon was observed in naturally infected tomatoes when a hot water dip (50 °C for 2 min) plus irradiation at 0.5 kGy totally eliminated *A. alternata* development under shelf-life conditions, but elevating the dose to 1 kGy resulted in 10% decay by *Alternaria*, although no external peel damage was observed (Barkai-Golan et al. 1991). In this case decay development can serve as a more sensitive criterion for unsuitable doses than can changes in the appearance of the fruit. As a matter of fact, Spalding

190 Case Studies

and Reeder (1986a) showed that while irradiation at 1 kGy markedly reduced soft rot in mature, green tomatoes inoculated with *Erwinia* spp. this treatment increased the incidence of fungal decay.

Enhanced decay development has frequently been reported for potato tubers at sprout-inhibiting doses of radiation. Susceptibility to rot varies with the variety but generally increases with the radiation dose, the extent of mechanical injuries, and duration of storage (Brownwell et al. 1957; Duncan et al. 1959; Sanyor and Dallyn 1961; El-Sayed and El-Wazeri 1977; Skou 1977; Matsuyama and Umeda 1983; Thomas 1983). Decay was also intensified in tubers with latent infections (Hooker and Duncan 1959; Farkas 1975). The enhanced rotting after irradiation has been attributed to the disturbance in metabolism, lack of wound periderm development, and decrease in natural resistance due to reduced synthesis of phytoalexins and phenolic compounds (Thomas 1983).

Similar to sprout-inhibiting chemicals (Cunningham 1953), irradiation may impair the formation of new periderm in wounds, thus providing easier access for microbial attack (Waggoner 1955; Skou 1977). Sparks and Iritani (1964) found that while irradiation of tubers soon after harvest considerably increased rotting, no difference in rot incidence was recorded when irradiation was applyed 70 days or more after harvest. It was suggested that tubers irradiated immediately after harvest may not have developed wound periderm by the time of irradiation. It is generally agreed that in order to avoid enhanced decay potato tubers should be given time and conditions for healing prior to irradiation but that irradiated tubers should also be protected against rewounding after irradiation (Skou 1977; Matsuyama and Umeda 1983). A reduced synthesis of the phytoalexins rishitin and albumin was reported by El-Sayed (1978c) in potato tubers at sprout-inhibiting doses of radiation. The increased incidence of rot in irradiated tubers was related, at least in part, to the reduction in the immune mechanism of the tubers as a result of the reduced accumulation of natural antifungal phytoalexins (El-Sayed 1978c; Langerak 1982).

Alteration of Postharvest Pathogen Populations. Alterations in the natural postharvest pathogenic flora have frequently been recorded in fruits and vegetables subjected to selective chemicals (Eckert and Ogawa 1985) or to certain formulae of controlled atmospheres (Barkai-Golan 1990). Several studies have reported the disruption of the balance between postharvest pathogens after irradiation at doses which inactivate radiation-sensitive pathogens but enable growth of resistant microorganisms. Beraha et al. (1959a) pointed out that doses which eradicated *P. digitatum* and *P. italicum* in citrus fruits allowed the development of *A. citri* at the stem-end of the fruit. This fungus, which is markedly more resistant to irradiation than the two *penicillia*, may be present at stem-end tissues of nonirradiated fruits as a latent infection. Irradiation at sublethal doses probably predisposes the fruit to active infection by this pathogen, by causing rind breakdown and tissue changes.

Irradiation at 1.2 kGy reduced the incidence of green mold on peeled oranges inoculated with *P. digitatum* spores (Barkai-Golan et al. 1984). However, this treatment allowed the development of *Cladosporium herbarum*, which is markedly more resistant to radiation than *P. digitatum*, and spores of which may readily infest the fruit surface but develop only after fruit peeling.

An increase in stem-end rot caused by *D. natalensis* and *P. citri* was reported by Spalding and von Windeguth (1988) in Tommy Atkins mangos irradiated at 150–750 Gy. A reduction in the incidence of the two fungi was, however, recorded at 1.5 kGy.

In sweet cherries, Cooper and Salunkhe (1963) found that while a dose of 3 kGy totally eliminated *Penicillium* decay, *Alternaria* and *Stemphylium* predominated on the irradiated fruit. These results are in accordance with the finding that the two multicellular spored fungi, *Alternaria* and *Stemphylium*, are more resistant to radiation than the unicellular spored *Penicillium* (Beraha et al. 1960; Salunkhe 1961).

Saravacos and Macris (1963) observed that the pathogen predominating on grape cultivars following irradiation at 1–2 kGy was *Pullularia pullulans* (*Aureobasidium pullulans*), which is exceptionally resistant to radiation. As a matter of fact, the high radioresistance of *Pullalaria* and the radiation-induced damage in the tissues, which facilitates fungal penetration, have been considered as the reasons for the increased in importance of this microorganism in various irradiated fruits and vegetables (Skou 1960, 1964a; Truelsen 1963).

Enhanced Mycotoxin Production. Mycotoxins, compounds toxic to man and animals, are secondary metabolites of fungi produced during fungal development on different media. *Aspergillus flavus* and *Aspergillus parasiticus*, both belonging to the *A. flavus* group (Raper and Fennell 1965), are mycotoxigenic species capable of producing aflatoxins, which are known for their hepatoxic and carcinogenic properties. *Aspergillus ochraceus* is another mycotoxigenic fungus and is capable of producing ochratoxin, which is a potent nephratoxin. These fungi are widespread on grains, seeds, peanuts, various nuts, almonds, spices, dried figs, and other food products (Christensen and Kaufmann 1965; Duggan 1970; Buchanan et al. 1975; Sommer et al. 1976). However, *A. flavus* has also been occasionally isolated from moldy tomatoes, grapes, and peaches while *A. ochraceus* was rarely isolated from stored grapes (Barkai-Golan 1981).

Several studies were carried out to determine the influence of radiation on the production of aflatoxin by *A. flavus* and *A. parasiticus*. Chang and Markakis (1982) reported a decreased aflatoxin production in barley with increase in the radiation dose at the range 0–4 kGy. Similar results were

192   Case Studies

observed by Ogbadu (1980), who pointed out the differences in aflatoxin yield on different media. However, several studies reported an increased aflatoxin production after spore irradiation (Jemmali and Guilbot 1970; Applegate and Chipley 1973a,b; Schindler et al. 1980). Applegate and Chipley (1973a,b, 1974b) found that while colony growth and sporulation of *A. flavus* in both wheat and synthetic media were greatly reduced after exposure of spores to irradiation at 3 kGy, higher yields of aflatoxins B_1, B_2, G_1, and G_2 were induced by 1.5–3 kGy doses. In addition to the dose and growth substrate, time of incubation after irradiation also affected aflatoxin accumulation (Applegate and Chipley 1974b).

The level of aflatoxin was similarly found to rise in some irradiated cereals, millets, and root vegetables inoculated with *A. parasiticus* spores (Priyadarshini and Tulpule 1976). An increase in the level of ochratoxin A was reported by Applegate and Chipley (1976) after exposing *A. ochraceus* spores to irradiation at 100–1000 Gy. Using mycelial discs as a source of inoculum, Paster et al. (1985) found that while irradiation at 1.5 and 2 kGy caused a 24 and 48 h delay in *A. ochraceus* colonization on a synthetic medium, doses of 0.5–2 kGy resulted in an increased amount of ochratoxin. No growth inhibition or notable change in ochratoxin production was recorded when mycelial discs were exposed to heat treatment (40 °C for 5 or 30 min) prior to irradiation. In contrast to mycelial discs, exposure of *A. ochraceus* sclerotia to gamma radiation or to combined treatments (heating at 60 °C for 15 or 30 min plus irradiation at 250 or 500 Gy), did not consistently result in different levels of ochratoxin (Paster and Barkai-Golan 1986).

Regarding mycotoxin production in fruits and vegetables the greatest concern has been for patulin production by *P. expansum*, the blue mold of pome and stone fruits (Buchanan et al. 1974; Sommer et al. 1974; Sommer and Buchanan 1978). Dose–response studies (Sommer et al. 1964a; Barkai-Golan et al. 1969b), showed that *P. expansum* is a radiosensitive pathogen: a dose of 1 kGy causes 99–100% spore inactivation while a dose of 2 kGy markedly delays colonization in vivo. However, no information is available on patulin accumulation following irradiation.

5.1.5 Conclusions

Complete sterilization of fresh fruits and vegetables by ionizing radiation is not feasible because of their sensitivity to doses which are lethal to the postharvest pathogen. Such doses may adversely affect the attributes of quality, such as appearance, aroma, flavor, and texture, and, in parallel, may enhance sensitivity to infection. However, for crops which are characterized by a short postharvest life, such as some cultivars of strawberries and raspberries, sublethal doses are beneficial in decay suppression by

Conclusions 193

halting pathogen growth temporarily. Other fruits may benefit from combinations of low radiation doses with other postharvest treatments, mainly heat or/and chemical application, due to synergistic or additive effects. As a component of such combinations, irradiation at doses below the threshold of damage to the host may lead to lethal effects which exclude the possibility of pathogen recovery.

As susceptibility of the fruit to infection generally increases with ripening, the ability of low radiation doses to retard the ripening process may also contribute indirectly to the delay in disease suppression by maintaining the natural resistance of the younger fruit to parasitism.

For most fruits and vegetables irradiation cannot substitute for postirradiation refrigerated storage (Maxie et al. 1971). However, irradiation was sometimes regarded as a complimentary treatment to low temperature storage (Baccaunaud 1988). Ripening inhibition at low doses has been reported for mango, papaya, banana, and other subtropical and tropical fruits. Since all these fruits are susceptible to chilling injury and cannot be held below 10–15°C (depending on cultivar and maturity stage) supplementary irradiation treatments to retard ripening may be useful (Kader 1986). Low temperature, in addition to extending the physiological life of fruits and vegetables by reducing their metabolic activities, may also retard decay development both directly, by retarding the growth of surviving microorganisms, and indirectly, by maintaining the resistance of the host to infection.

The difference in response to radiation of different cultivars of the same fruit or vegetable, or of a given cultivar from different growth environments or under different storage and transport conditions, emphasizes the need to determine the radiation treatment suitable for each case. For those commodities which benefit from single or combined radiation treatments, the delicate balance between the effective and the injurious dose level should, however, be carefully monitored.

The lack of protective residues in host tissues following irradiation has prompted the use of plastic films to prevent postirradiation infection and transit bruises. Such films, however, would be of little benefit in protecting the fruit against infections initiated and established in the field. With the development of new polymeric films, which enable the production of a modified atmosphere with the package (Barkai-Golan 1990), a combination of irradiation with the suitable wraps may not only prevent invasion by new pathogens but may also delay decay expression by latent infection due to the contribution of the modified atmosphere to ripening and senescence retardation.

Following extensive studies of "wholesomeness" of irradiated food during recent decades, the list of commodities approved for irradiation by the health authorities in various countries has been considerably lengthened. An important event took place in 1986 when gamma irradiation was ap-

194 Case Studies

proved by the Food and Drug Administration of the United States for treatment of fruits and vegetables up to a dosage of 1 kGy. This has accelerated both research and development in the field of ionizing radiation as a physical means for postharvest life extension.

5.2 Decontamination of Poultry Meat by Ionizing Radiation

I. Klinger[1] and M. Lapidot[2]

5.2.1 Microbiological Quality of Processed Poultry Meat

From the hygienic point of view, poultry meat has a special position among all other food products of animal origin. Modern poultry production, which is based upon mass raising, transportation, and industrial processing of poultry meat, acts as a homogenizer of microbiological contamination. The initial bacterial population on the skin of the live bird, originating from intestinal content and poultry house litter, is spread among individuals by common transportation and shared processing, which includes scalding, defeathering, evisceration, and immersion in chilled water, in the slaughter plant (Lillard 1989, 1990). In a study of the sources for cross-contamination of poultry during processing, it was concluded that all the processing plant operations, as summarized in Table 21, could affect the microbiological quality of the fully processed carcass (Bailey et al. 1987).

As a direct consequence of the extensive movements of chickens during processing a large variety of bacterial genera and species, including some pathogens, such as *Salmonella*, *Campylobacter*, *Staphylococcus*, *Clostridium*, and *Listeria*, can contaminate the meant (Table 22) (WAVFH 1967; Notermans and Kampelmacher 1974; Hobbs 1976; WHO 1976, 1982, 1984, 1988 a–c, 1989; Klinger et al. 1980, 1981; Blankenship 1986; Lahellec et al. 1986; Mossel 1987; Engel 1988; Fries 1987; Izat et al 1988; Beckers 1989).

From the practical viewpoint it means that a single flock, contaminated with one or more infective agents, can spread the infection to a whole production batch being processed. Around 60–80% of retail chickens in the UK are apparently contaminated with *Salmonella* and reports from other countries indicate levels which range from 5 to 73%. Up to 100% of the birds may contain *Campylobacter* and 60% may also harbor *Listeria monocytogenes* (Pini and Gilbert 1988).

---

[1] Kimron Veterinary Institute, P.O. Box 12, Bet Dagan 50250, Israel
[2] Soreq Nuclear Research Center, Yavneh, Israel

Microbiological Quality of Processed Poultry Meat

Table 21. Points of potential cross-contamination in poultry processing plants

Receiving and hanging: bird-to-bird in coops, air in holding sheds, coops, hands of hangers, dust and air in hanging area, shackles and rail dust

Killing: bird-to-bird, air, killing machine or knife, shackles and rail dust

Scalding and defeathering: scald water, picking fingers, condensate, air, bird-to-bird, pinners' hands, hock cutter, belt for rehang, shackles and rail dust, operators' hands

Evisceration: employees' hands, inspectors' hands, knives and other cutting instruments, machine contact surfaces (oil sac, lung machines, head cutters, etc.), air, shackles and rail dust, bird-to-bird, noncutting instruments (lung guns, lung rakes, head pullers, etc.), belts and chutes, giblet flumes and water, hang back rack

Chilling: chill water, ice, bird-to-bird, air, elevators, belts and chutes, giblet-to-giblet, neck-to-neck

Grading: employees' hands, belts, shackles and rail dust, bird-to-bird, air

Ice packing: employees' hands, packing bins, bird-to-bird, air, ice, packing material, giblet-or-neck-to-carcass (or vice versa)

Cup-up: employees' hands, saws or power knives, bird-to-bird, part-to-part, air, belts, bins, pans, shackles and rail dust

Table 22. Genera of bacteria isolated from poultry. (Cox and Bailey 1987)

Achromobacter	*Lactobacillus*
Actinomyces	*Listeria*
Aerobacter	*Microbacterium*
Alcaligenes	*Micrococcus*
Arthrobacter	*Neisseria*
Bacillus	*Paracolonbacterium*
Brevibacterium	*Proteus*
Campylobacter	*Providentia*
Clostridium	*Pseudomonas*
Corynebacterium	*Salmonella*
Escherichia	*Sarcina*
Flavobacterium	*Serratia*
Gyffkaya	*Staphylococcus*
Haemophilus	*Streptococcus*
	Streptomyces
	Yersinia

It is the goal of the poultry processor to reduce the total number of microorganisms and to prevent cross contamination of any pathogenic bacteria which might be present. The method commonly used for sanitizing poultry meat is rinsing in, or spraying with, large amounts of potable water, which reduces the bacterial load but only to a limited extent (Notermans and Kampelmacher 1974). Therefore, the contamination rate of poultry meat is relatively high compared to other foodstuffs of animal origin.

Although in most countries of the industrial world, chicken meat of excellent quality is processed, packaged, stored, and marketed under the best hygienic conditions and with good manufacturing practices, it is hardly possible to produce poultry meat free of pathogenic microorganisms. These can survive in processed meat or even multiply under refrigeration conditions, as demonstrated for *Listeria monocytogenes* (Engel 1988).

The presence of pathogenic bacteria on chicken meat creates a significant potential public health hazard. Although chicken meat is usually consumed after an exposure to heat treatment that is sufficient to destroy vegetative forms of pathogenic bacteria, secondary contamination, originating from raw chicken meat, occurs in food-processing establishments. Epidemiological studies of *Salmonella*, *Campylobacter*, and *Escherichia coli* have demonstrated that these microorganisms can survive on fingertips and other surfaces for varying periods of time and, in some cases, even after hand washing (WHO 1989). This is probably the most common reason for reporting chicken meat as the main source for outbreaks of diseases such as salmonellosis, campylobacteriosis, listeriosis, and other foodborne diseases.

Poultry carcasses leave the poultry slaughter plants with an unavoidable heterogeneous load of microorganisms on their surfaces. Large variations in the values of microbial loads of raw chicken meat, considered as "normal", may occur as a result of variations in processing technology, season of the year, and other factors (Klinger et al. 1980, 1981; Roberts 1990). The proportion of contaminated carcasses is determined mainly by the number of contaminated live birds entering the processing line and the hygienic conditions to which these carcasses are exposed. The number of pathogens per carcass is of little significance since, under favorable conditions, these pathogens multiply rapidly. For these reasons it was concluded that the definition of "absence of *Salmonella*" as a criterion for approval of poultry could not be justified owing to existing practical conditions. There is an immediate need for other effective measures, such as educational programs for proper handling and cooking of raw poultry, to reduce the possibility of cross contamination and consequent human infection. Such instructions are necessary, particularly in hospitals and geriatric institutions, where cooked poultry is included in therapeutic diets and is served to highly sensitive consumers. For the same reasons, poultry meat was considered by an expert committee as a food product with a relatively high initial bacterial load which leaves the processing line with a total bacterial count between $5 \times 10^5 – 1 \times 10^7$ counts/cm^2 (WHO 1989); this food product is highly perishable and can be spoiled easily.

5.2.2 Pathogenic Bacterial Contaminants of Poultry Meat

Salmonella. Although it was first recognized as a foodborne zoonotic agent more than 100 years ago, *Salmonella* still poses a severe problem in food hygiene and, in contrast to other zoonotic microorganisms which have been eradicated following preventive measures taken on the farm, the *Salmonella* problem is increasing (Silliker 1982; Mulder 1989). In fact, poultry has been recognized as the most important asymptomatic animal reservoir carrier and, consequently, excretor of *Salmonella* in the human food chain (D'Aust 1989). The contamination rate of *Salmonella* in poultry meat, even in the same country, varies considerably (Klinger et al. 1980; Silliker et al. 1980; Lahellec et al. 1986; Cox and Bailey 1987; Bailey 1988). It was demonstrated that the contamination rate in 15 federal inspected slaughter plants in the United States can vary from 2.5 to 73.7%, and from one year to another, between a reduction of 29.3 to an increase of 62.5% (Green et al. 1982). Paradoxically, the problem of poultry meat and egg contamination with *Salmonella* is severe in countries with a high level of sophisticated animal husbandry and where processing of products of animal origin occurs on fully industrialized mass production lines under the best hygienic conditions. It is estimated that 37% of the chicken meat harvested in the Western world is contaminated with salmonellae (WHO 1983, 1984). Concerning the contamination mechanism, it is clear that supply of eggs or chicks from infected parent breeder flocks could lead to a pyramidal increase in infection of progeny (D'Aust 1989). The limited number of parent breeder flock operators, and the varying degree of commitment in producing *Salmonella*-free stocks, has added to the complexity of the existing problem. Distribution of contaminated feeds to multiplier and broiler flocks, fecal contamination of water troughs in pens, use of new or old litter harboring *Salmonella*, and movement of insect and rodent vectors in barns further increase the potential for contamination of birds. Extensive serotypic profiles in vertically integrated poultry operations have clearly established a relationship between contamination in the broiler barn environment and the presence of *Salmonella* in finished products. Surface or internal contamination of hatching eggs and improper fumigation of eggs in the hatchery can lead to extensive cross contamination in the setting trays and to subsequent infection of newly hatched chicks. Shipment of birds from the broiler farm to the processing plant in poorly sanitized crates increases the bacterial load on feathers and the potential for cross contamination of birds and the slaughtering plant environment during processing. In the plant, the soiled birds are immersed in the scald tank, held at 52–60°C, to facilitate removal of feathers. Conditions prevailing in the scald tank are most favorable to *Salmonella* survival and surface inoculation of birds. Scalded birds are then introduced into a defeathering machine, a mechanical device that removes feathers through the beating action of rubber fingers. These rubber fingers effectively inoculate the surface of the birds that may have been *Salmonella*-free upon entering the processing plant. In addition, the possibility of carcass contamination during common evisceration and rinsing in chilling tanks further augments the risk of cross contamination. It was shown that bacteria are firmly attached to poultry skin before broilers arrive at the plant and that high numbers are still recovered after 40 consecutive whole carcass rinses of a single carcass (Lillard 1988). With prolonged immersion in a bacterial cell suspension, the attachment of *Salmonella* to chicken skin is related to an initial entrapment in a water film on the skin, followed by migration to the skin (Lillard 1989). This "adsorption" may introduce bacteria to carcasses that are inaccessible to bactericides. Baird Parker (1990), summing up the updated knowledge on foodborne salmonellosis, mentioned that despite attempts to reduce infection in live animals and to apply good hygienic practices in the abattoir and during further processing, it is inevitable that some raw meats will be contaminated. Ionizing radiation can eliminate the low numbers of *Salmonella* that may be present in raw meat. The advantage of irradiation is that it can be applied to the final packed product and, thus, recontamination is substantially avoided.

Campylobacter. Although *Campylobacter* live in a wide range of animals, they are especially common in birds, an adaptation which is reflected in their high optimum growth temperature of 42–43°C (Skirrow 1990). *Campylobacter jejuni* is a commensal bacterium of the intestinal tract of poultry. Intestinal contents of chickens colonized by *Campylobacter* may include as many as 5×10^3–1×10^7 *Campylobacter jejuni* per gram (Stern and Kazmi 1989). The problem is exacerbated by the ease with which contamination occurs during mass mechanized processing of poultry carcasses. During processing, although the number of *Campylobacter* on poultry carcasses is reduced by scald water, the organisms within the intestines remain viable and recontaminate the carcasses during defeathering and evisceration. The organisms are often isolated from poultry gizzards and livers, as well as from poultry meat. Freezing cannot entirely eliminate viable *Campylobacter*. Thus, the problem of contamination with *Campylobacter* shows great similarity to that with *Salmonella*. *Campylobacter* can be isolated from most poultry sold in the retail market. In one study fresh chickens had *Campylobacter* counts of 1.5×10^6 per bird and uneviscerated chickens had up to 2.4×10^7 counts per bird. A case-control study in Seattle in the United States attributed 48% of all cases of *Campylobacter* enteritis to the handling and consumption of chickens. Results are unlikely to be very different in Europe and other developed areas, where eating habits are similar to those in Seattle. A contamination rate with *Campylobacter* of 80.3% in chicken, 48% in duck, 38% in goose, and 3% in turkey was described by Kwiatek et al. (1990), and cross contamination, which takes place especially during defeathering, evisceration, and cooling of poultry by means of spin chillers, has been observed for poultry carcasses in the processing plants (WHO 1984; Lammerding et al. 1988). Apart from hygienic measures, the decontamination of end-products must be considered; to-date, treatment with lactic acid or irradiation of the meat offer the most promising results (Matches and Liston 1968; Kampelmacher 1983). *Campylobacter jejuni* isolated from poultry was found to be very sensitive to ionizing radiation (Lambert and Maxcy 1984).

Listeria. The contamination rate of raw chicken meat with *Listeria* and, particularly, with *Listeria monocytogenes*, a major foodborne pathogen, has recently received a great deal of attention (WHO 1988c; Johnson et al. 1990). A report on contamination of chicken meat with *Listeria monocytogenes* shows rates from 85% in the Federal Republic of Germany to 60% in England and Wales (Pini and Gilbert 1988). This parallels reports from the United States (Genigeorgis et al. 1989) on a contamination rate of 23% in retail broiler meat (Bailey et al. 1989) and of 70% in chicken parts. No epidemiological evidence on the relationship between the contamination of poultry and outbreaks of listeriosis in humans has been reported so far. The consumption of poultry meat usually follows its exposure to heat treatment, but it has also been demonstrated that *Listeria* spp. can survive, to some extent, the normal process of milk pasteurization (Huhtannen et al. 1989). The special characteristic of *Listeria* spp. of multiplying under refrigeration conditions and the survival of *Listeria* at heat treatments commonly used in the production of meat products (Harrison and Carpenter 1989a,b) render this microorganism particularly risky to the meat industry. Storage of industrially processed meat products, originating from poultry kept for several weeks under refrigeration conditions, may cause multiplication of residual *Listeria* bacteria, resulting in a heavy contamination of the product. It is, however, evident that irradiation of raw poultry meat at a dose of $2-2.5\,kGy$ can reduce the contamination rate by at least 1×10^4 counts per gram (Beckers 1989).

Under the conditions prevailing today it is not possible to produce pathogen-free chicken meat and, therefore, this product could be hazardous to public health. In is difficult to estimate the real costs of foodborne diseases to the economy of a particular country. Since public health authorities are usually poorly informed on the real incidence of foodborne diseases of microbial origin, this is often a hidden cost (WHO 1981a–c). Mossel (1977) estimates that only 1–5% of the actual outbreaks are even recorded and cases where only one or two members of a household are involved are usually not reported at all. This phenomenon, known as the "iceberg effect", means that most of the data are hidden. Foodborne diseases are considered to be second only to venereal diseases in causing morbidity in Europe. In the United States some 40 000 cases of human salmonellosis are officially reported annually by the Center of Disease Control (Bryan 1980), but their best estimate of cases is more like 2 million cases of illness and 2000 deaths annually. Skirrow (1990) estimates that the intangible costs of *Campylobacter* enteritis in England are £587 per patient; thus, the total costs of 32 000 laboratory diagnosed cases in 1989 amounted to nearly £9 million, and if the annual rate of 100 cases per 100000 persons is representative for the whole country the costs should be 10 times as high. Taking this factor into account, several attempts have been made to calculate the real costs of foodborne infections to a certain nation. Estimates of the real cost are increasing constantly. In the United States Eickhoff (1966) estimated the annual national losses due to salmonellosis in humans as between $10 and $100 million. Archer and Krenberg (1985) estimated the annual direct and indirect costs to public health in the United States due to contamination of food items with *Salmonella* as $2 billion. Roberts (1990), basing calculations on published estimates for medical costs and productivity losses associated with *Salmonella*, arrived at the value of $1.34 million and extrapolated costs for other specific bacteria causing foodborne infections, such as *Campylobacter*, to $1.47 billion. An annual cost of $480 million or more for listeriosis in the United States, reflecting 1860 cases calculated to occur each year, was estimated.

A listeriosis outbreak can be costly to the meat industry; the American meat industry loses annually 0.1–0.2% of a total income of $5.4 billions earned by 17 companies. The total costs of salmonellosis are far greater. The annual costs of human salmonellosis in 1977 in West Germany were estimated at DM 108 million for sickness and DM 12 million for death. The type of costs expressed as proportions of the social costs of salmonellosis in 1977 were: 42% through loss of leisure, 23% through welfare losses, 16% through treatment costs, 6% through examination costs, and 1% other costs. Krug and Rehm (1983) concluded that 72% of all *Salmonella* contamination of food originates from animals, and 81% of *Salmonella* transmitted to humans originates from food. Todd (1989a) estimated that the incidence of foodborne diseases in Canada, of microbial, parasitic, animal, plant, and chemical origin, reaches the figure of 2.2 million cases per annum, and is the causative agent for a possible 31.2 deaths for the same period of time, mainly resulting from *E. coli* hemorrhagic colitis, salmonellosis, and listeriosis. The total cost, including the value of deaths estimated to have occurred, came to $1.34 billion, of which 88% was caused by microbial diseases. It was suggested that this large sum of money for only one developed country could be reduced by various programs, such as education, good manufacturing and food-handling practices, appropriate sanitation procedures and, for certain diseases, like salmonellosis, for instance, irradiation of foods and competitive exclusion of pathogens in food animals. In another publication Todd (1989b) arrived at the estimation that, in the United States, there are annually 12.6 million cases of food-borne diseases, costing $8.4 billion. Microbial-borne diseases represent 84% of the United States costs, with salmonellosis and staphylococcal intoxications being the most economically important diseases (annually, $4.0 billion and $1.5 billion respectively). Other costly types of illness, which are transmitted to humans by poultry meat and are mentioned by

Todd (1989b), are listeriosis ($313 million), campylobacteriosis ($156 million), *Clostridium perfringens* enteritis ($123 million), and *E. coli* infections, including hemorrhagic colitis ($223 million). According to Baird Parker (1990), the estimated annual cost of salmonellosis in the United States is about $1.4 billion (about $700 per case). In 1986, estimated costs in the United Kingdom for treatment, diagnosis, and investigations were £375 per case. In North America costs per case ranged from about $800 for incidents associated with restaurants, take-away meals, and food prepared in the home, to about $10000 for cases associated with a food producer; costs for typhoid incidences were $10000 and $350000 respectively. Baird Parker (1990) concluded that the benefit of introducing measures that reduce or prevent the occurrence of salmonellosis in humans often outweighs the costs of installing preventive measures in the food-processing industry.

The contamination of food products with foodborne pathogens recently gained the healines in the popular mass media. The public is aware of the connection between health and food more than ever before. Information about the dangers of latent contamination of food items with pathogens is available to the public, which reacts by being more selective in its food-buying habits. The reaction of the public in the United Kingdom to the publicity given to the problem of *Salmonella* contamination of poultry and eggs, and to the contamination of cheeses with *Listeria*, was largely to reject these products. The impact of this phenomenon on the poultry industry during 1988 was more than severe and was described in some publications as "hysteria", with losses estimated at above £100 million.

5.2.3 Decontamination of Poultry Meat by Ionizing Radiation

The primary purpose of irradiation of poultry is to inactivate pathogenic microorganisms present in poultry meat and, thereby, to make these foods safer for human consumption. An additional benefit expected from this treatment is to extend the shelf-life of chilled poultry by reducing the microbial population, mainly vegetative forms of bacteria, which, due to their multiplication, cause finally the microbial spoilage of the product. The treatment is applied to the eviscerated poultry, to poultry parts, to edible offals, and to poultry products (minced meat, etc.) in the final retail packaging form to protect from recontamination. The treatment is applied to the chilled ($0-4\,^{\circ}$C) or to the frozen ($-18\,^{\circ}$C) product. Satisfactor decontamination of poultry meat using the ionizing radiation process has been a recommended solution for more than a decade (WAVFH 1967; WHO 1979, 1988a–c). Thayer et al. (1990) demonstrated for six *Salmonella* strains, artificially contaminating sterile, mechanically deboned chicken, that the D value needed was 0.56 kGy. Huhtannen et al. (1989) have shown

202 Case Studies

that a dose of 2 kGy was sufficient to destroy 1×10^4 cells of *Listeria monocytogenes*. Kampelmacher (1983) concluded that raising of *Salmonella*-free meat animals is not expected in the near future, since conventional decontamination processes are only partially efficient and are used so far only on a limited scale.

It has been shown that irradiation treatment can decontaminate poultry meat (WHO 1976; Urbain 1978a, 1983; Mulder 1982, 1989; Kampelmacher 1983; Klinger et al. 1986), and that the pathogenic bacteria, even at a high incidence, can easily be reduced to safe levels with a negligible or even unnoticed secondary effect on the organoleptic (Lee et al. 1985; Basker et al. 1986; Klinger et al. 1986) or nutritive properties of the product (Richardson 1955; Thomas and Calloway 1957; Raica et al. 1972). In addition, the total count of bacteria, consisting mainly of spoilage bacteria, is drastically reduced even if a low irradiation dose is employed, enabling prolongation of the preservation period of chilled chicken meat. Reports on the effect of ionizing radiation on microflora of poultry meat are summarized in Table 23.

The minimum absorbed dose which produces shelf-life extension of freshly chilled poultry and parts ranges from 1 to 2.50 kGy. In general, the smallest absorbed dose that is deemed effective under appropriate local conditions should be used. An excessively large absorbed dose may cause the formation of an "off flavor" in the poultry. This could occur above 2.5 (Sudarmadji and Urbain 1972) or 3.8 kGy (Camcigil 1990). The sensitivity to this "off flavor" formation varies with the source and species of the poultry. However, it may also depend very much on temperature, gaseous environment, type of part or organ (Camcigil 1990), and other factors. The thresholds mentioned should therefore be considered as indicative only. In addition, large absorbed doses may cause discoloration in some poultry. Care must, therefore, be taken on dose uniformity of poultry products treated in a certain size and shape of container and by certain types of irradiators. The minimum absorbed dose should be sufficient to achieve the technological purpose and the maximum should not exceed the tolerance limit of the product.

The minimum absorbed dose required to reduce the number of pathogenic bacteria, including *Salmonella*, *Campylobacter*, *Yersinia*, *Escherichia coli*, *Staphlylococcus*, and *Listeria*, present in frozen poultry meat and parts to levels commensurate with product safety for consumption, depends upon the initial level of contamination and the radiation sensitivity of the bacteria present. A precise absorbed dose, therefore, cannot be given without knowing the specific conditions. It is recommended, therefore, that the absorbed dose be determined for the conditions that exist locally (Klinger et al. 1986; WHO FAO/IAEA 1988). It seems that a minimum dose in the range of 3–5 kGy should be adequate for poultry irradiated in the frozen state. For this product, the maximum overall average dose

Decontamination of Poultry Meat by Ionizing Radiation

permitted by Codex and by some countries is 7 kGy. The threshold of "off flavor" formation is higher than in the case of freshly chilled poultry and may vary with source and species of the poultry as well as with other parameters. It must be determined locally in order to establish the permissible dose range.

Like any other food treatments, irradiation also has its limitations: it can inhibit multiplication of living microorganisms but it cannot destroy or denature metabolites, such as toxins (i.e., *Staphylococcus* enterotoxins), excreted by bacteria or the dead bacterial matrix. There was, therefore, a need to define bacteriological standards as "red border lines" for the maximal bacterial load of chicken to be admitted for decontamination by radiation. These values should by no means exceed the maximal load achieved in each country by those industrial slaughter plants which are equipped with suitable facilities, operate in accordance to the good manufacturing practice, and the controlled by the national regulatory authorities. The upper values of bacterial counts of ready-to-cook chilled or frozen chicken meat, prior to processing by the ionizing radiation technique, should be specified for each chicken product in accordance with the above considerations. An expert group (WHO/FAO/IAEA 1989) concluded that "food to be further processed should be of such microbiological quality that, were it not for the assumed presence of the target organisms, they would be considered wholesome and microbiologically acceptable." This was done to give guidance to national authorities that this harmless, non-detectable, and effective treatment should not be misused, and chicken meat with inferior microbiological quality, or even unfit for human consumption, due to a heavy bacterial load prior to the irradiation process, should not be offered to the public as a fresh, clean, safe, and wholesome product after cleaning by exposure to the irradiation process.

It is important to notice that not every microorganism which is harbored in chicken meat is affected similarly by ionizing radiation. Due to differences in the D_{10} value for different bacteria the overall effect varies. Small differences in sensitivity were noticed even in the same species (Mulder 1982). Most species found on chicken meat are very sensitive to ionizing radiation. However, because of the selective effect of the radiation process the surviving bacteria, which consist of those resistant to the doses employed, will multiply in particular if they also possess some psychrophilic properties. This will eventually limit the shelf-life of the chilled irradiated product. Tables 24 and 25 describe the relative sensitivity of various bacteria to the irradiation treatment and the D value needed for elimination of various microorganisms present on chicken meat.

Spoilage microorganisms in chicken meat will quickly render it unacceptable due to putrefacation and decomposition of the product, resulting in unpleasant odors and development of slime. These effects are slowed down by low initial loads of microorganisms and low temperatures during storage.

204   Case Studies

Table 23. Investigations on the effects of ionizing radiation on the contaminating microflora and properties of chicken meat

Year	Author	Irradiation dose (kGy)	Effects
1968	Idziak and Incze	5	Extended shelf-life at 5 °C (14 days). 10–11 log reduction in the number of viable *Salmonella* and *Staphyloccocus*. Shift in the microbial population (isolation of microorganisms tentatively identified as *Moraxella* and *Herellea*)
1969	Idziak and Whitaker	3.5	Simplification in the microbial ecology pattern when compared to nonirradiated controls
1974	Bakalivanov et al.	1–3	1 kGy was almost as effective as 3 kGy in prolonging storage life to 3 weeks; control carcasses became unacceptable by the 10th day
1975	Gruenewald	2.5–8	75 deep-frozen broilers were irradiated and examined for sensory changes during 2 years storage at −30 °C. No significant differences in color, flavor, smell, consistency, and overall quality between irradiated and nonirradiated broilers were found.
1977	Mulder et al.	2.5	Considerable reduction in the total number of *Salmonella* on contaminated broiler carcass; 2.5 kGy reduced naturally occurring salmonellae by 2.5 cycles
1980	Fiszer et al.	2.5–5	Little effect of irradiation on fat-quality indices was observed. Differences in sensory properties of the irradiated broiler meat were moderate. Total plate count decreased considerably as a result of irradiation, then increased gradually during storage
1982	Kiss and Farkas	2–5	Two- to threefold extension of the keeping quality of chicken meat without noticeable deterioration of organoleptic quality. Elimination of *Salmonella* infection of the carcasses
1982	Mulder	2.5	Reduction of Enterobacteriaceae counts to low levels and considerable reduction in the incidence of *Salmonella* (none were detected after 1 month at 5 °C or 4 months at −18 °C)
1984	Bok and Halzapfel	3–7	Irradiation eliminated *Salmonella* and other pathogens; sensory characteristics of irradiated chickens were good. At 4 °C, shelf-life was 3 for nonirradiated, 13 for irradiated at 3 kGy, and more than 30 days for chickens irradiated at 7 kGy
1985	Lee et al.	5–10	Sam gei tang prepared from irradiated chicken

Table 23. *Continued*

Year	Author	Irradiation dose (kGy)	Effects
			and stored for 15 days was similar to that prepared from fresh chickens. Steamed chicken was superior when prepared from irradiated 15-day-old samples
1985	Cho et al.	5–10	Gamma irradiation could extend the shelf-life of chicken meat by 2–4 weeks
1986	Basker et al.	3.7	The eating quality of leg meat was satisfactory for at least 1 week and decreased after about 3 weeks. Breast meat was satisfactory for about 3 weeks; decrease in quality was noticed after about 4 weeks of storage
1986	Klinger et al.	2–4.5	Chicken became free of salmonellae, coliforms, and staphylococci and had reduced bacterial counts. Following chill storage for about 4 weeks, irradiation resistant *Moraxella* spp. (D = 0.83 kGy) predominated. Extensive taste panel tests showed that sensory quality of meat immediately after irradiation with about 3.7 kGy did not differ from that of control samples. Sensory differences developed on chilled storage; the sensory quality of irradiated meat decreased to an unacceptable level over about 3–4 weeks
1986	El-Husseiny et al.	1–5	The number of spoilage microorganisms increased progressively during chilled storage, but the rates of increase in the irradiated samples were lower than in the untreated controls. The effect of irradiation on decreasing the count of pathogens was lower in frozen than in chilled storage. No growth of pathogens was observed after treatment with 3 or 5 kGy
1989	Hanis et al.	0.5–10	*P. auroginosa* was eliminated by 1.0 kGy, *Serratia marescens* by 2.5–50 kGy, and *Salmonella typhimurium* by 10.0 kGy. The characteristic radiation odor, which increased with the radiation dose and temperature, was removed by heating during meat preparation. Radiation increased acid and peroxide values and partially destroyed thiamin and riboflavin. Less decreases in fat indexes and less destruction of vitamins were observed after irradiation at lower temperatures; amino acid content was not affected by the treatment

206 Case Studies

Table 24. Effect of irradiation on the number of microorganisms ($\log_{10}$ CFU/g) on frozen chicken. (Mossel 1987)

Microorganism	Irradiation (kGy)				
	0	1	2	3	4
Mesophilic colony count	6.8	5.8	4.6	4.1	3.6
Psychrotrophic colony count	5.8	5.7	4.0	<2.8	<1.8
Enterobacteriaceae	5.5	<2.8	1.0	0.4	−0.4
Lactobacillus	6.0	4.1	4.2	3.1	<2.8
Lancefield D streptococci	5.1	3.7	3.9	3.2	<2.0
Staphylococcus aureus	4.6	2.2	<−0.5	<−0.5	<−0.5

CFU, colony forming units.

Table 25. Effect of irradiation on the percentage of total colony count of mesophilic and psychrotrophic fractions on frozen chicken. (Mossel 1987)

Type of organism	Microorganism	Irradiation (kGy)				
		Mesophilic			Psychro-trophic	
		0	2	4	0	2
Gram-positive cocci	*Aerococcus*	–	10	–	–	–
	Micrococcus	28	39	43	–	83
	Staphylococcus	10	–	–	–	–
	Lancefield D streptococci	3	43	50	–	–
Gram-positive rods	*Corynebacterium*	19	3	–	19	3
	Lactobacillus	22	–	–	–	–
Gram-negative rods	*Acinetobacter*	2	–	–	8	–
	Xanthomonas	–	–	–	4	–
	Pseudomonas	–	–	–	46	–
	Kluyvera	–	–	–	5	–
	Hafnia	10	–	–	7	–
	Klebsiella	–	–	–	11	–
	Escherichia coli	4	–	–	–	–
Yeasts		2	5	7	–	14

Limit of significance, 2–5%.

Storage of chicken meat under refrigeration conditions has an effect of selection on the residual flora of the product (Table 26).

It was demonstrated by Mossel (1987) that 90% of the bacterial population on chicken meat carcasses belongs to the *Pseudomonas* species, which can grow under refrigeration conditions and are extremely sensitive toward the irradiation process. Inactivation of these bacteria by the irradi-

Table 26. Flora shift (%) as a result of postirradiation storage at 21 °C for 36 h (severe temperature abuse) and 12 °C for 84 h (moderate temperature abuse).[a] (Mossel 1987)

Type of organism	Organism	Irradiation (kGy)			
		Mesophilic		Psychrotrophic	
		0	4	0	2
Gram-positive cocci	*Micrococcus*	13 (9)	– (–)	5 (2)	– (–)
	Staphylococcus	2 (14)	– (–)	3 (–)	– (–)
Gram-positive rods	*Corynebacterium*	23 (34)	4 (–)	22 (23)	– (–)
	Lactobacillus	13 (9)	– (–)	13 (6)	– (–)
	Leuconostoc	4	–	–	–
Gram-negative rods	*Moraxella*	17 (25)	72 (93)	30 (46)	62 (88)
	Pseudomonas	4 (–)	– (7)	3 (2)	2 (12)
	Enterobacteriaceae	6 (–)	– (–)	3 (8)	– (–)
	Acinetobacter	4	24	18	36
Other		14 (9)	– (–)	3 (13)	– (–)

[a] Values for moderate temperature abuse are given in parentheses.

ation process can slow down the spoilage process and therefore prolong shelf-life, resulting in an increased marketing period. Storage of radiation-treated chicken under refrigeration conditions has also an effect of selection on the residual bacterial flora which resists the irradiation process. This depends on several factors, such as packaging, which prevents recontamination of the processed products, storage conditions, particularly the storage temperature, and the irradiation dose employed. Chilled storage of irradiation-treated chicken may result in spoilage of the product if it is stored for longer periods of time; this is due to multiplication of the irradiation-resistant and psychrophilic species, such as *Moraxella* spp., which are present in an insignificant number on the product (Idziak and Incze 1968; Idziak and Whitaker 1969; Silliker et al. 1980; Klinger et al. 1986).

Foods that have been treated with quantities of ionizing energy up to 10 kGy are not sterile and dependence must be placed upon other methods of preservation to prevent multiplication of surviving microorganisms. Where microorganisms survive, investigations have been concerned about the possibilities of:

1. development of resistance to ionizing energy in the surviving organisms;
2. increased virulence of pathogens;
3. unusual spoilage characteristics due to changes in the normal flora; and

208   Case Studies

4. changes in physiological characteristics that would make it difficult to identify the organisms.

To-date, however, there is no evidence to indicate that any of these possibilities are valid. As far is known, none of these four possibilities present a risk to the consumer (WHO 1976, 1982; Ingram and Farkas 1977; Maxcy 1983). Consideration of early data (1963–1973) indicated that doses of 2–5 kGy reduced the numbers of salmonellae and of other pathogens by 4–7 logs, those of clostridia by 1–2 logs, those of *Achromobacter* and fecal streptococci to a few survivors, and those of *Pseudomonas* to almost nil (Thornley 1963; Licciardello 1970; Maxcy 1983; Mossel 1987). No occurrences of unusual pathogens were reported, so that the overall result was considered as a marked improvement in microbiological status related to food poisoning. No microbiological problems were considered to arise in the case of storage of irradiated frozen chicken. In the case of storage of chilled chicken, *Pseudomonas* developed predominantly at the lower doses and lower storage temperatures, whereas *Achromobacter* developed at higher doses and higher storage temperatures. No special problems were likely to arise because of their similarity to the normal spoilage flora. As regards the *Moraxella* bacteria, which are less sensitive toward the irradiation process and are present in irradiated chilled chickens, they were also found in considerable numbers on normal chickens and were supposed to be of no special significance (Kampelmacher 1977). In order to prevent multiplication of eventual survivors of the *Salmonella* group, a temperature not exceeding 5–6 °C was recommended (Michener and Elliot 1964; Matches and Liston 1968). To protect against surviving spores of mesophilic clostridia, such as *Clostridium botulinum* types A and B, a storage temperature below 10 °C was considered to be sufficient (Matches and Liston 1968). Accordingly, the Codex Alimentarius Commission concluded that there are no microbiological safety problems with moist foods, such as fresh poultry that has been treated with medium doses (up to 10 kGy) of ionizing energy, as long as they are stored and distributed near the temperature of ice (2–5 °C) and according to good manufacturing practice (FAO/WHO 1982). It is considered that in the unlikely occurrence of contamination of fresh poultry with *Clostridium botulinum* type E the product would be safe at a practical commercial dose treatment of up to 3 kGy, as permitted in the United States FDA (FDA 1990). The surviving members of the natural microflora would be able to multiply at 10 °C and would produce spoilage odors within 8 days, whereas the *Clostridium botulinum* type E survivors could not produce toxin within 14 days. At higher temperatures, even at 30 °C, the other surviving microflora would grow and produce spoilage before *Clostridium botulinum* type E toxin was produced (Firstenberg-Edln et al. 1982). This spoilage would be an adequate warning to prevent consumption of the product (CAST 1986).

References

Abdel-Kader AS, Morris LL, Maxie EC (1968) Physiological studies of gamma irradiated tomato fruits II. Effects on deterioration and shelf-life. Proc Am Soc Hortic Sci 93: 831–842

Acuff GR, Vanderzant C, Hanna MO, Ehlers JG, Gardner FA (1986) Effects of handling and preparation of turkey products on the survival of *Campylobacter jejuni*. J Food Prot 49: 627–631

Aharoni Y, Barkai-Golan R (1987) Pre-harvest fungicide sprays and polyvinyl wraps to control *Botrytis* rot and prolong the post-harvest life of strawberries. J Hortic Sci 62: 177–181

Ahmed AA, Watrous GH, Hargrove GL, Dimick PS (1976) Effects of fluorescent light on flavor and ascorbic acid content in refrigerated orange juice and drinks. J Milk Food Technol 39: 332–336

Ahmed EM, Dennison RA, Merkley MS (1968) Effects of low level irradiation upon the preservation of food products. University of Florida, Dept Food Sci, Annu Rep ORO-675. US Atomic Energy Commission, Washington DC, pp 9–12

Ahmed ES (1977) Biochemical response of skin-coated citrus fruits irradiated for preservation. Arab J Nucl Sci Appl 10: 155–158

Akamine EK, Goo T (1977a) Effects of gamma irradiation on shelf-life extension of fresh papaya (*Carica papaya* L. var. Solo). Hawaii Agric Exp Stn Bull No 165

Akamine EK, Goo T (1977b) Effects of gamma irradiation on shelf-life of fresh lychees (*Lichi chinensis* Sonn.). Hawaii Agric Exp Stn Bull No 169

Akamine EK, Moy JH (1983) Delay in postharvest ripening and senesence of fruits. In: Josephson ES, Peterson MS (eds) Preservation of food by ionizing radiation, vol III. CRC Press, Boca Raton, pp 129–158

Akamine EK, Wong R (1965) Shelf-life extension of papayas with gamma irradiation. In: Dosimetry, tolerance, and shelf-life extension related to disinfestation of fuits and vegetables by gamma irradiation. Rep 1964–1965, Div Isot Dev, US Atomic Energy Commission, Washington DC, pp 55–60

Alabastro EF, Pineda AS, Pangan AC, del Valle MJ (1978) Irradiation of fresh Cavendish bananas (*Musa cavendishi*) and mangoes (*Mangifera indica* Linn. var *carabao*). The microbiological aspect. In: Food preservation by irradiation, vol I. International Atomic Energy Agency, Vienna, pp 283–303

Aleixo JAG, Swaminathan B, Jamesen KS, Pratt DE (1985) Destruction of pathogenic bacteria in turkeys roasted in microwave ovens. J Food Sci 50: 873–875

Allen C, Parks OW (1979) Photodegradation of riboflavin in milks exposed to fluorescent light. J Dairy Sci 62: 1377–1379

Alper T, Gillies NE (1958) "Restoration" of *Escherichia coli* strain B after irradiation: its dependance on suboptimal growth conditions. J Gen Microbiol 18: 461–472

Ammerman G, Adres C (1973) New equipment/techniques produce high quality seafood products. Food Process 34: 76–78

210 References

Andrews J, Atkinson GF (1984) Modification of a microwave oven for laboratory use. J Chem Educ 61: 177–178

Annis PJ (1980) Design and use of domestic microwave ovens. J Food Prot 43: 629–632

Anonymous (1977) Manual of food irradiation dosimetry, Tech Rep Ser 178, IAEC, Vienna

Anonymous (1983) Radiation preservation of foods. Food Technol 37: 55–60

Applegate KL, Chipley JR (1973a) Increased aflatoxin production by *Aspergillus flavus* via cobalt irradiation. Poult Sci 52: 1492–1496

Applegate KL, Chipley JR (1973b) Increased aflatoxin G_1 production by *Aspergillus flavus* via irradiation. Mycologia 65: 1266–1273

Applegate KL, Chipley JR (1974a) Daily variation in the production of aflatoxins by *Aspergillus flavus* NRRL 3145 following exposure to ^{60}Co irradiation. J Appl Bacteriol 37: 359–372

Applegate KL, Chipley JR (1974b) Effects of ^{60}Co gamma irradiation on aflatoxin B_1 and B_2 production by *Aspergillus flavus*. Mycologia 66: 436–445

Applegate KL, Chipley JR (1976) Production of ochratoxin A by *Aspergillus ochraceus* NRRL-3174 before and after exposure to ^{60}Co irradiation. Appl Environ Microbiol 31: 349–353

Arakawa O, Hori Y, Ogata R (1985) Relative effectiveness and interaction of ultraviolet-B, red and blue light in anthocyanine synthesis of apple fruit. Physiol Plant 64: 323–327

Archer DL, Krenberg JE (1985) Incidence and costs of foodborne diarrhoea disease in the United States. J Food Prot 48: 887–894

Aref MM, Brach EJ, Tape NW (1969) A pilot-plant continuous-process microwave oven. Can Inst Food Technol J 2: 37–41

Attoe EL, von Elbe JH (1981) Photochemical degradation of betanine and selected anthocyanins. J Food Sci 46: 1934–1937

Atwood KC, Pittenberg TH (1955) The relation between the X-ray survival curves of *Neurospora* microconidia and ascospores. Genetics 40: 563–564

Avisse C, Varoquaux P (1977) Microwave blanching of peaches. J Microwave Power 12: 73–79

Ayoub JA, Berkowitz D, Kenyon EM, Wadsworth CK (1974) Continuous microwave sterilization of meat in flexible pouches. J Food Sci 39: 309–313

Baccaunaud M (1988) Ionizing. A complementary treatment to the use of low temperature. Inf Centre Technique Interprofessionel de Fruits et Legumes, France, No. 45, pp 21–26

Bailey JS (1988) Status and prospects for control of *Salmonella* contamination of poultry. Poult Sci 67: 920–947

Bailey JS, Thomson JA, Cox NA (1987) Contamination of poultry during processing. In: Cunningham FE, Cox NA (eds) The microbiology of poultry meat products, Academic Press, New York, pp 193–206

Bailey JS, Fletcher DL, Cox NA (1989) Recovery and serotype distribution of *Listeria monocytogenes* from broiler chickens in the southeastern United States. J Food Prot 52: 148–150

Bailey ME, Frame RW, Nauman HD (1964) Cured meat pigments-studies of the photooxidation of nitrosomyoglobin. Agric Food Chem 12: 89–93

Baird Parker AC (1990) Foodborne salmonellosis. Lancet 336: 1231–1235

Bakalivanov S, Nikolova T, Mitkov S (1974) Prolongation of storage life of refrigerated fresh poultry meat by gamma irradiation. Khranitelna Prom 24: 26–28

Bakanowski SM, Zoller JM (1984) Endpoint temperature distributions in microwave and conventionally cooked pork. Food Technol 38: 45–51

References 211

Baker CW, Doly NC (1977) Microwave conditioning of durum wheat. II. Optimization of semolina yield and spaghetti quality. J Agric Food Chem 25: 819–822

Bala K, Nauman HD (1977) Effect of light on the color stability of sterile aqueous beef extract. J Food Sci 42: 563–564

Baldwin RE, Fields ML, Poon WC, Korschgen B (1971) Destruction of salmonellae by microwave heating of fish with implications for fish products. J Milk Food Technol 34: 467–470

Barbeau WE, Schnepf M (1989) Sensory attributes and thiamine content of roasting chickens cooked in a microwave, convection microwave and conventional electric oven. J Food Qual 12: 203–213

Barkai-Golan R (1971) Production of pectolytic and cellulolytic enzymes in irradiated culture of *Penicillium digitatum* Sacc. Radiat Bot 11: 215–218

Barkai-Golan R (1981) An annotated check list of fungi causing postharvest diseases of fruits and vegetables in Israel. Division of Scientific Publications, Spec Publ 194, ARO, The Volcani Center, Bet Dagan

Barkai-Golan R (1985) Technological development on food irradiation in Israel over the past two decades. SAFFOST '85 Cong, Pretoria, vol 2, pp 342–358

Barkai-Golan R (1990) Postharvest disease suppression by atmospheric modifications. In: Calderon M, Barkai-Golan R (eds) Food preservation by modified atmospheres, CRC Press, Boca Raton, pp 237–264

Barkai-Golan R, Kahan RS (1966) Effects of gamma radiation on extending the storage life of oranges. Plant Dis Rep 50: 874–877

Barkai-Golan R, Kahan RS (1967) Combined action of diphenyl and gamma radiation on the in vitro development of fungi pathogenic to citrus fruits. Phytopathology 58: 696–698

Barkai-Golan R, Kahan RS (1968) The effect of gamma irradiation on stored grapes and on the in vitro development of fungi and yeasts causing grape rots. Prelim Rep No 605, The Volcani Inst Agric Res, Bet Dagan

Barkai-Golan R, Kahan RS (1971) The effect of radiation on the pathogenicity of fungi and yeasts causing rot in stored grapes. Isr J Agric Res 21: 4

Barkai-Golan R, Karadavid R (1991) Cellulolytic activity of *Penicillium digitatum* and *P. italicum* related to fungal growth and to pathogenesis in citrus fruits. J Phytopathol 13: 65–72

Barkai-Golan R, Padova R (1970a) Combined irradiation, heat and biphenyl treatments for the control of *Penicillium digitatum* in inoculated citrus fruits. Isr J Agric Res 20: 129–132

Barkai-Golan R, Padova R (1970b) Attempts to reduce post irradiation damage to the peel of stored field-infected citrus fruit. Rep IA-1218, Isr Atomic Energy Commission, pp 175–176

Barkai-Golan R, Padova R (1975) Extending the shelf-life of cultivated mushrooms by gamma radiation. Hassadeh 56: 113–117 (Hebrew with English summary)

Barkai-Golan R, Padova R (1981) Eradication of *Penicillium* on citrus fruits by electron radiation. Proc Int Soc Citric 2: 799–801

Barkai-Golan R, Kahan RS, Lattar FS (1966) A white sporeless mutant of *Penicillium digitatum* obtained after irradiation of *Penicillium digitatum* Sacc. Isr J Agric Res 16: 95–96

Barkai-Golan R, Temkin-Gorodeiski N, Kahan RS (1967) Effect of gamma irradiation on the development of *Botrytis cinerea* and *Rhizopus nigricans*, causing rot in strawberry fruits. Food Irradiat 8: 34–36

Barkai-Golan R, Kahan RS, Temkin-Gorodeiski N (1968) Sensitivity of stored melon fruit fungi to gamma irradiation. Int J Appl Radiat Isot 19: 579–583

212 References

Barkai-Golan R, Kahan RS, Padova R (1969a) Synergistic effects of gamma radiation and heat on the development of *Penicillium digitatum* in vitro and in stored citrus fruits. Phytopathology 59: 922–924

Barkai-Golan R, Ben-Arie R, Guelfat-Reich S, Kahan RS (1969b) Sensitivity to gamma radiation of fungi pathogenic to pears. Int J Appl Radiat Isot 20: 577–583

Barkai-Golan R, Ben-Yehoshua S, Aharoni N (1971) The development of *Botrytis cinerea* in irradiated strawberries during storage. Int J Appl Radiat Isot 22: 155–158

Barkai-Golan R, Kahan RS, Padova R, Ben-Arie R (1977) Combined treatment of heat or chemicals with radiation to control decay in stored fruits. In: Biological science. Proc Worksh on the Use of ionizing radiation in agriculture, Wageningen, EUR 5815 EN, pp 157–169

Barkai-Golan R, Karadavid R, Padova R (1984) Control of *Penicillium* rot on peeled citrus fruit by gamma radiation and plastic wrappers. Proc Int Soc Citric 1: 506–507

Barkai-Golan R, Lavy-Meir G, Kopeliovitch E (1989) Effects of ethylene on the susceptibility to *Botrytis cinerea* infection of different tomato genotypes. Ann Appl Biol 114: 391–396

Barmore CR, Brown GE (1985) Influence of ethylene on increased susceptibility of oranges to *Diplodia natalensis*. Plant Dis 69: 228–230

Bartholomew BP, Ogden LV (1990) Effect of emulsifiers and fortification methods on light stability of vitamin A in milk. J Dairy Sci 73: 1485–1488

Bartz JA (1980) Causes of postharvest losses in a Florida tomato shipment. Plant Dis 64: 934–937

Basker D, Klinger I, Lapidot M, Eisenberg E (1986) Effect of chilled storage of radiation pasteurized chicken carcasses on the eating quality of the resultant cooked meat. J Food Technol 21: 437–41

Beckers HJ (1989) The occurrence of *Listeria* in food. In: Kampelmacher EH (ed) Foodborne listeriosis – Symposium held on September 7, 1988 in Wiesbaden, FRG. Behrs, Hamburg, pp 85–97

Beelman R (1988) Factors influencing post harvest quality and shelf-life of fresh mushrooms. Mushroom J 182: 455–463

Beer A (1852) Bestimmung der Absorption des rothen Lichts in farbigen Flüssigkeiten. Ann Physik 86: 78–88

Belli-Donini ML, Pansolli P (1970) Gamma irradiation of table grapes of the "Hoanes" variety. Food Irradiat 10: 15–21

Ben-Arie R, Barkai-Golan R (1969) Combined heat-radiation treatments to control storage rots of Spadona pears. Int J Appl Radiat Isot 20: 687–690

Bengtsson NE, Risman PO (1971) Dielectric properties of foots at 3 GHz as determined by a cavity perturbation technique. II. Measurements of food materials. J Microwave Power 6: 107–113

Bengtsson NE, Ohlsson T (1974) Microwave heating in the food industry. Proc IEEE 62: 44–47

Bengtsson NE, Kheen W, Del Valle FR (1970) Radiofrequency pasteurization of cured hams. J Food Sci 35: 681–687

Beraha L (1964) Influence of radiation dose rate on decay of citrus, pears, peaches, and on *Penicillium italicum* and *Botrytis cinerea* in vitro. Phytopathology 54: 755–759

Beraha L, Ramsey GB, Smith MA, Wright WR (1959a) Factors influencing the use of gamma radiation to control decay of lemons and oranges. Phytopathology 49: 91–96

References

213

Beraha L, Ramsey GB, Smith MA, Wright WR (1959b) Effects of gamma radiation on brown rot and *Rhizopus* rot of peaches and the causal organisms. Phytopathology 49: 354–356

Beraha L, Ramsey GB, Smith MA, Wright WR (1959c) Effects of gamma radiation on some important potato tuber decays. Am Potato J 36: 333–338

Beraha L, Smith MA, Ramsey GB, Wright WR (1960) Gamma radiation dose response of some decay pathogens. Phytopathology 50: 474–476

Beraha L, Ramsey GB, Smith MA, Wright WR, Heiligman F (1961) Gamma radiation in the control of decay in strawberries, prapes and apples. Food Technol 15: 94–98

Beraha L, Garber ED, Stromaes O (1964) Genetics of phytopathogenic fungi. Virulence of color and nutritionally deficient mutants of *Penicillium italicum* and *Penicillium digitatum*. Can J Bot 42: 429–436

Berg PT, Marchello MJ, Erickson DO, Slanger WD (1985) Selected nutrient content of beef longissimus muscle relative to marbling class, fat status, and cooking method. J Food Sci 50: 1029–1033

Berk S (1953) The effects of ionizing radiation from Polonium on the spores of *Aspergillus niger*. Mycologia 45: 488–506

Berry BW, Leddy K (1984) Beef patty composition: effects of fat content and cooking method. J Am Diet Assoc 84: 654–658

Betts RP, Farr L, Bankes P, Stringer MF (1988) The detection of irradiated foods using the direct epifluorescent filter technique. J Appl Bacteriol 64: 329–335

Bhartia P, Stuchly SS, Hamid MAK (1973) Experimental results for combinational microwave and hot air drying. J Microwave Power 8: 245–248

Bialod D, Jolion M, LeGoff R (1978) Microwave thawing of food products using associated surface cooling. J Microwave Power 13: 269–274

Biela AM, McGill AEJ (1985) Can baby feeding equipment be sterilised in the domestic microwave oven? JR Soc Health 105: 131–132

Bird RP, Draper HH (1984) Comparative studies on different methods of malonaldehyde determination. Methods Enzymol 105: 184–191

Birth GS (1978) The light scattering properties of foods. J Food Sci 43: 916–925

Bishop C (1988) There's no fraud like an old fraud. New Sci 115: 52–55

Bishop RC, Klein RM (1975) Photo-promotion of anthocyanin synthesis in harvested apples. Hortic Sci 10: 126–127

Blanco JF, Dawson LE (1974) Survival of *Clostridium perfringens* on chicken cooked with microwave energy. Poult Sci 53: 1823–1830

Blankenship R (1986) Reduction of spoilage and pathogenic bacteria. Broiler Ind: 24–25

Boag TS, Johnson GI, Izard M, Murray C, Fitzsimmons KC (1990) Physiological responses of mangoes cv. Kensington Pride to gamma irradiation treatment as affected by fruit maturity and ripeness. Ann Appl Biol 116: 177–187

Bogl W, Heide L (1985) Chemiluminescence measurements as an identification method for gamma-irradiated foodstuffs. Radiat Phys Chem 25: 173–185

Bogunowic M, Katusic-Razem B, Ivekovic V (1986) Improvement of strawberry storability by ionizing radiation. Jugosl Vocarstvo 20: 659–664

Bok HE, Holzapfel WH (1984) Extension of shelflife of refrigerated chicken carcasses by radurization. Food Rev 11: 69–71

Bolin HR, Nury FS, Bloch F (1964) Effect of light on processed dried fruits. Food Technol 18: 1975–1976

Booker JL, Friese MA (1989) Safety of microwave-interactive paperboard packaging materials. Food Technol 43: 110–118

214 References

Bosset JO, Daget N, Desarzens C, Dieffenbacher A, Fluckiger E, Lavancy P, Nick B, Pauchard JP, Tagliaferri E (1986) The influence of light transmittance and gas permeability of various packing materials on the quality of whole natural yoghurt during storage. Lebensm Wiss Technol 19: 104–106

Bothill DE, Hawker JS (1970) Chlorophylls and their derivatives during drying of sultana grapes. J Sci Food Agric 21: 193–196

Bouguer P (1760) Traité d'optique sur la graduation de la lumière, Paris

Boulanger RJ, Boerner WM, Hamid MAK (1969) Comparison of microwave and dielectric heating systems for the control of moisture content and insect infestations of grain. J Microwave Power 4: 194–197

Bouno MA, Niroomand F, Fung DYC, Erickson LE (1989) Destruction of indigenous *Bacillus* spores in soymilk by heat. J Food Prot 52: 825–826

Boyd DR, Crone AVJ, Hamilton JTG, Hand MV, Stevenson MH, Stevenson PJ (1991) Synthesis, characterization and potential use of 2-dodecylcyclobutanone as a marker for irradiated chicken. J Agric Food Chem 39: 789–792

Brady PL, Haughey PE, Rothschild MF (1985) Microwave and conventional heating effects on sensory quality and thiamin content of flounder and haddock fillets. Home Econ Res J 14: 236–240

Bramlage WJ, Couey HM (1965) Gamma radiation of fruits to extend market life. Market Res Rep 717, Agricultural Research Service, US Department of Agriculture, Washington DC

Bramlage WJ, Lipton WJ (1965) Gamma radiation of vegetables to extend market life. Marketing Res Rep No. 703, Agricultural Research Service, US Department of Agriculture, Washington DC

Bray SL, Duthie AH, Rogers RP (1977) Consumers can detect light-induced flavor in milk, J Food Prot 40: 586–587

Bridges BA, Horne T (1959) The influence of environmental factors on the microbicidal effect of ionizing radiation. J Appl Bacteriol 22: 96–115

Brodrick HT (1982) Ways in which irradiation could improve the effectiveness of several existing preservation processes in South Africa. S Afr Food Rev 9: 29–32

Brodrick HT, Strydom GJ (1984) The radurisation of bananas under commercial conditions. Part 1. Increased storage life. Citrus Sub-Trop Fruit J 602: 4–6

Brodrick HT, Thomas AC (1978) Radiation preservation of subtropical fruits in South Africa. In: Food preservation by irradiation, vol I. International Atomic Energy Agency, Vienna, pp 167–178

Brodrick HT, van der Linde HJ (1981) Technological feasibility studies on combination treatments for subtropical fruits. In: Combination processes in food irradiation. International Atomic Energy Agency, Vienna, pp 141–152

Brodrick HT, van der Westhuisen GCAA (1976) Identification of the causal organism of mango soft brown rot. Phytophylactica 8: 13–16

Brodrick HT, Jacobs CJ, Kok IB, Milne DL, Basson RA, Thomas AC (1972) Application of ionizing radiation to the preservation of subtropical fruit in South Africa. Proc Natl Conf Technol Appl Nucl Tech, Atomic Energy Board, Pelindaba, Republic of South Africa

Brodrick HT, Thomas AC, Visser F, Beyers M (1976) Studies on the use of gamma irradiation and hot water treatments for shelf life extension of papayas. Plant Dis Rep 60: 749–753

Brodrick HT, Thomas AC, van Tonder AJ, Terblanche JC (1977) Combined heat and gamma-irradiation treatments for the control of strawberry disease under market conditions. Atomic Energy Board, Pelindaba, Republic of South Africa, PER-7, 17

References 215

Brodrick HT, Thord-Gray RS, Strydom GJ (1984) Radurisation of bananas under commercial conditions, part 2. Shelf – or market-life extension. Citrus Sub-Trop Fruit J 603: 4–5

Brodrick HT, Thord-Gray RS, Strydom GJ (1985) Post-harvest disease control of plums and nectarines with radurisation treatment. SAFFOST '85 Congr, Pretoria, vol 2, pp 391–398

Brower HJ, Cline LD (1984) Response of *Trichogramma pretiosum* and *T. evanescens* to whitelight, blacklight or no-light suction traps. Fl Entomol 67: 262–269

Brown GH, Morrison WC (1954) An exploration of the effects of strong radio-frequency fields on microorganisms in aqueous solution. Food Technol 8: 361–366

Brownwell NE (1952) Utilization of gross fission products. Eng Res Inst Univ of Michigan, Prog Rep, Ann Arbor 1951–1952

Brownwell NE, Gustafson FG, Nehemias JB, Isleib DR, Hooker WJ (1957) Storage properties of gamma-irradiated potatoes. Food Technol 11: 306–312

Bruce WA (1975) Effects of UV radiation on egg hatch of *Plodia interpunctella* (Lepidoptera: Pyralidae). J Stored Prod Res 11: 243–248

Bryan FL (1980) Foodborne diseases in the United States associated with meat and poulty. J Food Prot 43: 140–150

Buchanan JR, Sommer NF, Fortlage RJ, Maxie E, Mitchell FG, Hsieh DPH (1974) Patulin from *Penicillium expansum* in stone fruits and pears. J Am Soc Hortic Sci 99: 262–265

Buchanan JR, Sommer NF, Fortlage RJ (1975) *Aspergillus flavus* infection and aflatoxin production in fig fruits. Appl Microbiol 30: 238–241

Buckley P, Sommer NF, Gortz JH, Maxie EC (1967) Effects of chemical protection on repair of potentially lethal irradiation injury in *Rhizopus stolonifer* sporangiospores. Radiat Res 30: 275–282

Bunch WL, Matthews ME, Marth EH (1976) Hospital chill food service systems: acceptability and microbiological characteristics of beef-soy loaves when processed according to system procedures. J Food Sci 41: 1273–1276

Camcigil M (1990) Regulating trade in irradiated food. Spec Rep to the Director General, International Atomic Energy Agency, Vienna

Campbell JD, Stothers S, Vaisey M, Berk B (1968) Gamma irradiation influence on the storage and nutritional quality of mushrooms. J Food Sci 33: 540–542

Carlin F, Zimmermann W, Sundberg A (1982) Destruction of trichina larvae in beef-pork loaves cooked in microwave ovens. J Food Sci 47: 1096–1099

Carnevale J, Cole ER, Crank G (1980) Photocatalyzed oxidation of paprika pigments. J Agric Food Chem 18: 953–956

Carrol DE, Lopez A (1969) Lethality of radio-frequency upon microorganism in liquid, buffered and alcoholic food systems. J Food Sci 34: 320–324

Carroll LE (1989) Hydrocolloid functions to improve stability of microwavable foods. Food Technol 43: 96–100

CAST (1986) Ionizing energy in food processing and pest control. I. Wholesomeness of food treated with ionizing energy, Council for Agricultural Science and Technology Rep 109

Castle L, Jickells SM, Sharman M, Gramshaw JW, Gillbert J (1988a) Migration of the plasticizer acetyltributyl citrate from plastic film into foods during microwave cooking and other domestic use. J Food Prot 51: 916–919

Castle L, Mercer AJ, Gilbert J (1988b) Migration from plasticized films into foods. 4. Use of polymeric plasticizers and lower levels of di-(2-ethylhexyl)adipate

216 References

plasticizer in PVC films to reduce migration into foods. Food Addit Contam 5: 277–282

Castle L, Mayo A, Crews C, Gilbert J (1989) Migration of poly(ethylene terephthalate) (PET) oligomers from PET plastics into foods during microwave and conventional cooking and into bottled beverages. J Food Prot 52: 337–342

Castle L, Mayo A, Gilbert J (1990) Migration of epoxidised soya bean oil into foods from retail packaging materials and from plasticised PVC film used in the home. Food Addit Contam 7: 29–36

Chachin K, Minamide T, Ogata K (1974) Effect of cathode rays on phenols content and some enzyme activities of *Citrus unshiu*. Food Irradiat Jpn 9: 69–73

Chalutz E, Maxie EC, Sommer NF (1965) Interaction of gamma irradiation and controlled atmospheres on *Botrytis* rot of strawberry fruit. Proc Am Soc Hortic Sci 88: 365–371

Chan HWS (1977) Photosensitized oxidation of unsaturated fatty acid methyl esters. the identification of different pathways. J Am Oil Chem Soc 54: 100–102

Chan HWS, Levett G, Griffiths NH (1977) Light-induced flavor deterioration. The effect of exposure to light of pork luncheon meat containing erythrosine. J Sci Food Agric 28: 339–342

Chan HWS, Levett G, Griffiths NH (1978) Light-induced flavor deterioration. Exposure of potato crisps to light and its effect on subsequent storage in the dark. J Sci Food Agric 29: 1055–1060

Chang HG, Markakis P (1982) Effect of gamma irradiation on aflatoxin production in barley. J Sci Food Agric 33: 559–564

Chen TC, Culotta JT, Wang WS (1973) Effects of water and microwave energy pre-cooking on microbiological quality of chicken parts. J Food Sci 38: 155–157

Chicoye E, Powrie WD, Fennema O (1968) Photooxidation of cholesterol in spray-dried egg yolk upon irradiation. J Food Sci 33: 581–584

Chin HB, Kimball JR Jr, Hung J, Allen B (1985) Microwave oven drying determination of total solids in processed tomato products: collaborative study. J Assoc Off Anal Chem 68: 1081–1083

Chipley JR (1980) Effects of microwave irradiation on microorganisms. Adv Appl Microbiol 26: 129–145

Cho HO, Lee MK, Byun MW, Kwon JH, Kim JG (1985) Radurization of the microorganisms contaminating chicken. Korean J Food Sci Technol 17: 170–174

Chowdhury SU, Hamid MA (1969) Some aspects of the rate of appearance and type of microbial damage in irradiated bananas. Food Irradiat 10: 18–24

Christensen CM, Kaufmann HH (1965) Deterioration of stored grains by fungi. Annu Rev Phytopathol 3: 69–84

Coggle JE (1973) Biological effects of radiation. Wykeham Publications, London

Cooper GM, Salunkhe DK (1963) Effect of gamma-radiation, chemical and packaging treatments on refrigerated life of strawberries and sweet cherries. Food Technol 17: 123–126

Copson DA (1975) Microwave heating. Avi, Westport, CT

Costello CA, Morris WC, Barwick JR (1990) Effects of heating bacon and sausage in nonwoven, melt-blown material. J Food Sci 55: 298–300

Cotter DJ, Sawyer RL (1961) The effect of gamma irradiation on the incidence of black spot, and ascorbic acid, glutathione and tyrosinase content of potato tubers. Am Potato J 38: 58–65

Couey HM, Bramlage WJ (1965) Effect of spore population and age of infection on the response of *Botrytis cinerea* to gamma radiation. Phytopathology 55: 1013–1015

References

Cox NA, Bailey JS (1987) Pathogens associated with processed poultry. In: Cunningham FE, Cox NA (eds) The microbiology of poultry meat products. Academic Press, New York, pp 293–316

Craven SE, Lillard HS (1974) Effect of microwave heating of pre-cooking chicken on *Clostridium perfringens*. J Food Sci 39: 211–212

Cremer ML, Richman DK (1987) Sensory quality of turkey breasts and energy consumption for roasting in a convection oven and reheating in infrared, microwave, and convection ovens. J Food Sci 52: 846–850

Crespo FL, Ockermann HW (1977) Thermal destruction of microorganisms in meat by micrwoave and conventional cooking. J Food Prot 40: 422–444

Crespo FL, Ockermann HW, Irvin KM (1977) Effect of conventional and microwave heating on *Pseudomonas putrefaciens, Streptococcus faecalis* and *Lactobacillus plantarum* in meat tissue. J Food Prot 40: 588–591

Culkin KA, Fung DYC (1975) Destruction of *Escherichia coli* and *Salmonella typhimurium* in microwave-cooked soups. J Milk Food Technol 38: 8–15

Cunningham FE (1980) Influence of microwave radiation on psychrotrophic bacteria. J Food Prot 42: 651–655

Cunningham HS (1953) A historical study of the influence of sprout inhibitors on *Fusarium* infection of potato tubers. Phytopathology 43: 95–98

Curnutte B (1980) Principles of microwave radiation. J Food Prot 43: 618–624

Dahl CA, Matthews ME, Marth EH (1981) Survival of *Streptococcus faecium* in beef loaf and potatoes after microwave heating in a simulated cook/chill hospital food service system. J Food Prot 44: 128–133

Dallyn SL, Sawyer RL (1954) Effect of sprout inhibiting levels of gamma irradiation on the quality of onions. Proc Am Soc Hortic Sci 73: 398–406

D'Aust JY (1989) Salmonella. In: Doyle MP (ed) Foodborne bacterial pathogens. Marcel Dekker, New York, pp 327–446

Davidson I, Forrester AR (1988) γ-Irradiation of spices. In: Rice-Evans C, Dormandy T (eds) Free radicals: chemistry, pathology and medicine. Richelieu Press, London, pp 271–292

Davis BD, Dulbecco R, Eisen HN, Ginsberg HS, Wood WB Jr (1970) Microbiology. Harper and Row, New York

Davis CE, Searcy GK, Blankenship LC, Townsend WE (1988) Pyruvate kinase activity as an indicator of temperature attained during cooking of cured pork. J Food Prot 51: 773–777

Day PR (1957) Mutation to virulence in *Cladosporium fulvum*. Nature (Lond) 179: 1141–1142

Dealler SF, Lacey RW (1990) Superficial microwave heating. Nature (Lond) 344: 496

Decareau RV (1986) Microwave food processing equipment throughout the world. Food Technol 40: 99–105

Decareau RV, Peterson RA (1986) Microwave processing and engineering. VCH, Weinheim

DeMan JM (1981) Light-induced destruction of vitamin A in milk. J Dairy Sci 64: 2031–2032

Den Drijver L, Holzapfel CW, van der Linde HJ (1986) High-performance liquid chromatography determination of D-arabinohexos-2-ulose (D-glucosone) in irradiated mango. J Agric Food Chem 34: 758–762

Dennis C (1983) Soft fruits. In: Dennis C (ed) Postharvest pathology of fruits and vegetables Academic Press, London, pp 23–42

Dennison RA, Ahmed EM (1966) Review of the status of irradiation effects on citrus fruits. In: Proc Int Symp Food irradiation, International Atomic Energy Agency, Vienna, pp 619-634

Dennison RA, Ahmed EM (1971) Effects of low-level irradiation on the preservation of fruits: a 7-year summary. Isot Radiat Technol 9: 194-200

Dennison RA, Ahmed EM (1975) Irradiation treatment of fruits and vegetables. In: Haard NF, Salunkhe DK (eds) Postharvest biology and handling of fruits and vegetables. Avi Westport, CT, pp 118-129

Desrosiers MF (1989) Gamma-irradiated seafoods: identification and dosimetry by electron paramagnetic resonance spectroscopy. J Agric Food Chem 37: 96-100

Desrosiers MF (1990) Assessing radiation dose to food. Nature (Lond) 345: 485

Desrosiers MF, McLauglin WL (1989) Examination of gamma-irradiated fruits and vegetables by electron spin resonance spectroscopy. Radiat Phys Chem 34: 895-898

Desrosiers MF, Simic MG (1988) Postirradiation dosimetry of meat by electron spin resonance spectroscopy of bones. J Agric Food Chem 36: 601-603

Dhaliwal AS, Salunkhe DK (1963) Ionizing radiation and packaging effects on respiratory behaviour, fungal growth, and storage-life of peaches *Prunus persica*. Radiat Bot 3: 75

Dharkar SD, Sreenivasan A (1966) Irradiation of tropical fruits and vegetables. In: Proc Int Symp on Food irradiation, International Atomic Energy Agency, Vienna, pp 635-650

Dharkar SD, Savagaon KA, Srirangarajan AN, Sreenivasan A (1966a) Irradiation of mangoes. I. Radiation-induced delay in ripening of Alphonso mangoes. J Food Sci 31: 863-869

Dharkar SD, Savagaon KA, Srirangarajan AN, Sreenivasan A (1966b) Irradiation of mangoes. II. Radiation effects on skin-coated Alphonso mangoes. J Food Sci 31: 870-877

Diehl JF (1972) Elektronenspinresonanz-Untersuchungen an strahlenkonservierten Lebensmitteln, II. Einfluß des Wassergehaltes auf die Spinkonzentration. Lebensm Wiss Technol 5: 51-53

Dietrich WC, Huxsoll CC, Wagner JR, Guadagni DG (1970) Comparison of microwave with steam or water blanching of corn-on-the cob. 2. Peroxidase inactivation and flavor retention. Food Technol 24: 293-296

Dillard CJ, Tappel AL (1984) Fluorescent damage products of lipid peroxidation. Methods Enzymol 105: 337-341

Dodd NJF, Lea JS, Swallow AJ (1988) ESR detection of irradiated food. Nature (Lond) 334: 387

Dodd NJF, Lea JS, Swallow AJ (1989) The ESR detection of irradiated food. Int J Appl Radiat Isot 40: 1211-1214

Dohmaru T, Furuka M, Katayama T, Toratani H, Takeda A (1989) Identification of irradiated pepper with the level of hydrogen gas as a probe. Radiat Res 120: 552-555

Doleiden FH, Fahrenholtz SR, Lamola AA, Trozzolo AM (1974) Reactivity of cholesterol and some fatty acids toward singlet oxygen. Photochem Photobiol 20: 519-521

Doty NC, Baker CW (1977) Microwave conditioning of hard red spring wheat. I. Effects of wide power range on flour and bread quality. Cereal Chem 54: 717-727

Draganic IG, Draganic ZD (1971) The radiation chemistry of water. Academic Press, New York

References 219

Droby S, Prusky D, Jacoby B, Goldman A (1986) Presence of antifungal compounds in the peel of mango fruits and their relation to latent infections of *Alternaria alternata*. Physiol Mol Plant Pathol 29: 173–183

Dubery IA, Schabort JC (1987) 6,7-Dimethoxycoumarin: a stress metabolite with antifungal activity in γ-irradiated citrus peel. S Afr J Sci 83: 440–441

Dubery IA, Holzapfel CW, Kruger GJ, Schabort JC, van Dyk M (1988) Characterization of a γ-radiation-induced antifungal stress metabolite in citrus peel. Phytochemistry 27: 2769–2772

Duggan RE (1970) Controlling aflatoxins. FDA Pap Aprill 1970, Washington DC, pp 13–18

Duncan DT, Hooker WJ, Heiligman F (1959) Storage rot susceptibility of potato tubers exposed to minimum sprout inhibiting levels of ionizing radiation. Food Technol 13: 159–164

Du Venage CA (1985) Strawberry radurisation on a commercial scale. SAFFOST '85 Congr, Pretoria, vol 2, pp 463–467

Eckert JW (1978) Pathological diseases of fresh fruits and vegetables. In: Hultin HO, Milner N (eds) Postharvest biology and biotechnology. Food and Nutrition Press, Westport, CT, pp 161–209

Eckert JW, Ogawa JM (1985) The chemical control of postharvest diseases of subtropical and tropical fruits. Annu Rev Phytopathol 23: 421–445

Eckert JW and Sommer NE (1967) Control of diseases of fruits and vegetables by postharvest treatment. Annu Rev Phytopathol 5: 391–432

Ehlermann D (1972) The possible identification of an irradiation treatment of fish by means of electrical (ac) resistance measurement. J Food Sci 37: 501–504

Eickhoff TC (1966) Economic losses due to salmonellosis in humans. In: Lineweaver H (ed) The destruction of *Salmonella*. Western Regional Research Laboratory USDA, Albany, CA, pp 2195–2227

El-Husseiny TM, El-Fouly MZ, Neweigy NA, El-Mongy TM (1986) The effect of gamma radiation on decreasing the spoilage microorganisms in chilled and frozen chicken carcasses. Ann Agric Sci Ain Shams Univ 31: 937–948

Elias PS, Cohen AJ Eds (1977) Radiation chemistry of major food components. Elsevier, Amsterdam

Elias PS, Cohen AJ Eds (1983) Recent advances in food irradiation. Elsevier, Amsterdam

El-Kazzaz MK, Sommer NF, Fortlage RJ (1983) Effect of different atmospheres on postharvest decay and quality of fresh strawberries. Phytopathology 73: 282–285

El-Sayed SA (1978a) Changes in keeping quality of tomato fruits after postharvest treatment with gamma irradiation combined with heat. Egypt J Hortic 5: 167–174

El-Sayed SA (1978b) Control of post-harvest storage decay of soft-type date fruits with special reference to the effect of gamma irradiation. Egypt J Hortic 5: 175–182

El-Sayed SA (1978c) Phytoalexins as possible controlling agents of microbial spoilage of irradiated fresh fruits and vegetables during storage. In: Food preservation by irradiation vol 1. International Atomic Energy Agency, Vienna, pp 179–193

El-Sayed SA, El-Wazeri SM (1977) Application of an induced phytoalexin formed by pepper fruits to control rot incidence in irradiated potatoes. Egypt J Hortic 4: 157–163

El-Zawahry YA, Rowley DB (1979) Radiation resistance and injury of *Yersenia enterocolitica*. Appl Environ Microbiol 37: 50–54

220 References

Engel RE (1988) Foodborne listeriosis: risk from meat and poultry. Paper presented at the WHO Working Group on Foodborne listeriosis, Geneva, Switzerland, February 15–19

Engel RE, Post AR, Post RC (1988) Implementation of irradiation of pork for trichina control. Food Technol 42: 71–75

Esaka M, Suzuki K, Kubota K (1987) Effects of microwave heating on lipoxygenase and trypsin inhibitor activities, and water absorption of winged bean seeds. J Food Sci 50: 1738–1739

Esterbauer H, Zollner H (1989) Methods for determination of aldehydic lipid peroxidation products. Free Rad Biol Med 7: 197–203

Esterbauer H, Lang J, Zadravec S, Slater TS (1984) Detection of malonaldehyde by high-performance liquid chromatography. Methods Enzymol 105: 319–328

Ettinger KV, Puite KJ (1982) Lyoluminescence dosimetry. In: McLaughlin WL (ed) Trends in radiation dosimetry. Pergamon Press, Oxford

Fanslow GE, Saul RA (1971) Drying field corn with microwave power and unheated air. J Microwave Power 6: 229–233

FAO/WHO (1982) The microbiological safety of irradiated food. Report of the Board of the Int Comm Food Microbiol Hygiene (ICFMH) of the Int Union Microbiol Soc (IUMS) Copenhagen CAC, Rome, CX/FH 83/9

Farkas J (1975) Present status and prospects for the commercialization in Hungary of irradiated food items for human consumption. In: Requirements for the irradiation of food on a commercial scale. International Atomic Energy Agency, Vienna, pp 37–59

Farkas J, Koncz A, Sharif MM (1990) Identification of irradiated dry ingradients on the basis of starch damage. Radiat Phys Chem 35: 324–328

Farooqui WA, Ahmad M, Hussain A, Hussain AM (1974a) Effects of gamma radiation on mangoes (*Mangifera indica* L.) stored under different conditions. J Agric Res Pak 12: 31–42

Farooqui WA, Ahmad M, Hussain A, Naqvi MH (1974b) Effect of gamma irradiation on Kinnow Mandarin during storage. Nucleus (Karachi) 11: 25–31

FDA (1986) Irradiation in the production, processing and handling of food; Final rule (21 CFR Part 179), Federal Register 51(75): 13376–13399 (April 18, 1986)

FDA (1990) Irradiation in the production processing and handling of food; Final rule (21 CFR Part 179), Federal Register 55(85) 18538–18544 (May 2, 1990)

Fedeli E, Brillo A (1975) Meccanismo dei fenomeni di autossidazione delle sostanze grasse. Nota II: Olio d'oliva. Riv Ital Sostanze Grasse 52: 109–113

Ferguson WE, Yates AR (1966) The effects of gamma radiation on bananas. J Food Technol 20: 105–107

Finney EE (1978) Engineering techniques for nondestructive quality evaluation of agricultural products. J Food Prod 41: 57–62

Firstenberg-Eden R, Rowley DB, Shattuck GE (1982) Factors affecting growth and toxin production by *Clostridium botulinum* type E on irradiated (0.3 Mrad) chicken skin. J Food Sci 47: 867–870

Fiszer W, Zabielski J, Mroz J (1980) Radurization of poultry meat. In: Proc 26th Eur Meet of Meat research workers, Colorado Springs, USA, August 31– September 5 1980, vol 1, pp 248–251

Flor HH (1958) Mutation to wider virulence in *Melampsora lini*. Phytopathology 48: 297–301

Foley J, O'Donovan D, Cooney C (1971) Photocatalyzed oxidation of butter. J Soc Dairy Technol 24: 38–45

Foley J, Gleeson JJ, King JJ (1977) Influence of pasteurization and homogenization treatments on photocatalyzed oxidation of cream. J Food Prot 40: 25–28

References   221

Foote CS (1976) Photosensitized oxidation and singlet oxygen: consequences in biological systems. In: Pryor WA (ed) Free radicals in biology, vol 2. Academic Press, New York, pp 85–133

Foote CS, Clough RL, Tee BG (1978) Photooxidation of tocopherols. In: deDuve, Hayaishi O (eds) Tocopherol, oxygen and biomembranes. Elsevier, Amsterdam, pp 13–25

Frankel EN (1980) Lipid oxidation. Prog Lipid Res 19: 1–22

Frankel EN (1985) Chemistry of free radical and singlet oxidation of lipids. Prog Lipid Res 23: 197–221

French J (1653) The art of distillation. T Williams, London, p 42

Fries R (1987) Qualitative/quantitative Untersuchungen auf *Salmonella* bei industriell gewonnenen Geflügelfleisch. Dtsch Tieraerztl Wochenschr 94: 197–200

Frimer AA (1985) Singlet O_2. CRC Press, Boca Raton

Fritsch G, Reymond D (1970) Effects of X-rays on pectin studied by electron spin resonance. J Appl Radiat Isot 21: 329–334

Fruin JT, Guthertz LS (1982) Survival of bacteria in food cooked by microwave oven, conventional oven and slow cookers. J Food Sci 45: 695–698

Fung DYC, Cunningham FE (1980) Effect of microwaves on microorganisms in foods. J Food Prot 43: 641–650

Gardiol FE (1984) Introduction to microwaves. Artech House, Dedham, MA

Gaylord AM, Warthesen JJ, Smith DE (1986) Influence of milk fat, milk solids, and light intensity on the light stability of vitamin A and riboflavin in lowfat milk. J Dairy Sci 69: 2779–2784

Genigeorgis CA, Dutulescu D, Garayzabal JF (1989) Prevalence of *Listeria* spp. in poultry meat at the supermarket and slaughterhouse level. J Food Prot 52: 618–624

Georgiev I (1983) The effect of gamma irradiation in conjunction with the fungicide captan on suppressing grey rot during storage of cv. "Bolgar" grapes. Gradinar Lozar Nauka 20: 103–108

Georgopoulos SG, Macris B, Georgiadou E (1966) Reduction of radiation resistance in fruit spoilage fungi by chemicals. Phytopathology 56: 230–234

Gerling JE (1986) Microwaves in the food industry: promise and reality. Food Technol 40: 82–83

Gesner C (1599) The practise of the new and old phisicke. P Short, London, p 23–24

Gilmore TM, Dimik PS (1979) Photochemical changes in major whey proteins of cow's milk. J Dairy Sci 62: 189–194

Goksu-Ogelman HY, Regulla DF (1989) Detection of irradiated food. Nature (Lond) 340: 23

Gordy W, Pruden B, Snipes W (1965) Some radiation effects on DNA and its constituents. Proc Natl Acad Sci USA 53: 751–755

Gray R, Stevenson MH (1989a) The effect of post-irradiation cooking on the ESR signal in irradiated chicken drumsticks. Int J Food Technol 24: 447–450

Gray R, Stevenson MH (1989b) Detection of irradiated deboned turkey meat using electron spin resonance spectroscopy. Radiat Phys Chem 34: 899–902

Gray R, Stevenson MH, Kilpatrick DJ (1990) The effect of irradiation dose and age of bird on the ESR signal in irradiated chicken drumsticks. Radiat Phys Chem 35: 284–287

Grecz N, Walter AA, Anellis A (1964) Effect of radio-frequency energy (2450 MHz) on bacterial spores. Bacteriol Proc 1964: 145

222 References

Grecz N, Rowley DB, Matsuyama A (1983) The action of radiation on bacteria and viruses. In: Josephson ES, Peterson MS (eds) Preservation of food by ionizing radiation, vol II. CRC Press, Boca Raton, pp 167–218

Green SS, Moran AB, Johnston RW, Uhler P, Chiu J (1982) The incidence of *Salmonella* species and serotypes in young whole chicken carcasses in 1979 as compared with 1967. Poult Sci 61: 288–293

Grierson W, Dennison RA (1965) Irradiation treatment of 'Valencia' oranges and 'March' grapefruit. Proc Fla State Hortic Soc 78: 233–237

Grootveld M, Jain R (1989a) Recent advances in the development of a diagnostic test for irradiated foodstuffs. Free Rad Res Commun 6: 271–292

Grootveld M, Jain R (1989b) Methods for the detection of irradiated foodstuffs: aromatic hydroxylation and degradation of polyunsaturated fatty acids. Radiat Phys Chem 34: 925–934

Grootveld M, Jain R, Claxson AWD, Naughton D, Blake DR (1990) The detection of irradiated foodstuffs. Trends Food Sci Technol 1: 7–14

Gruenewald T (1975) Lagerversuch mit bestrahlten tiefgefrorenen Hähnchen. Fleischwirtschaft 55: 1093–1095

Gunasekaran S (1990) Delayed light emission as a means of quality evaluation of fruits and vegetables. CRC Crit Rev Food Sci Nutr 29: 19–34

Hafez YS, Mohamed AI, Hewedy FM, Singh G (1985) Effects of microwave heating on solubility, digestibility and metabolism of soy protein. J Food Sci 50: 415–417

Hafez YS, Mohamed AI, Perera PA, Singh G, Hussein AS (1989) Effects of microwave heating and gamma irradiation on phytate and phospholipid contents of soybean (*Glycine max* L.). J Food Sci 54: 958–962

Hamid MAK, Boulanger RJ (1969) A new method for the control of moisture and insect infestations of grain by microwave power. J Microwave Power 4: 11–16

Hamid MAK, Kshyap CS, van Cauwenberghe R (1968) Control of grain insects by microwave power. J Microwave Power 3: 126–131

Hamid MAK, Boulanger RJ, Long SC, Gallop RA, Pereira RR (1969) Microwave pasteurization of raw milk. J Microwave Power 4: 272–275

Hamid MAK, Mostowy NJ, Bhartia P (1975) Microwave bean roaster. J Microwave Power 10: 109–112

Hamrick PE, Butler BT (1973) Exposure of bacteria to 2450 MHz microwave radiation. J Microwave Power 8: 227–233

Hanis T, Jelen P, Klir P, Mnukova J, Perez B, Pesesk M (1989) Poultry meat irradiation, effect of temperature on chemical changes and inactivation of micro-organisms. J Food Prot 52: 26–29

Harrison DL (1980) Microwave versus conventional cooking methods. Effects on food quality attributes. J Food Prot 43: 633–637

Harrison MA, Carpenter SL (1989a) Survival of *Listeria monocytogenes* on micro-wave cooked poultry. Food Microbiol 6: 153–157

Harrison MA, Carpenter SL (1989b) Survival of large populations of *Listeria monocytogenes* on chicken breasts processed using moisture heat. J Food Prot 52: 376–378

Hart EJ (1972) Radiation chemistry of aqueous solutions. Radiat Res Rev 3: 285–304

Hart RJ, White JA, Reid WJ (1988) Technical note: occurrence of o-tyrosine in non-irradiated foods. Int Food Sci Technol 23: 643–647

Hashimoto T, Yatsuhashi H (1984) Ultraviolet photoreceptors and their interaction in broom sorghum. Analysis of action spectra and fluence-response curves. In:

References 223

Senger H (ed) Blue light effects in biological systems. Springer, Berlin Heidelberg New York, pp 125–136

Hashisaka AE, Matches JR, Batters Y, Hungate FP, Dong FM (1990) Effects of gamma irradiation at −78 °C on microbial populations in dairy products. J Food Sci 55: 1284–1289

Hatton TT Jr, Beraha L, Wright WR (1961) Preliminary trials of gamma radiation on mature-green Irwin and Sensation mangos. Proc Fla Mango Forum, 21st Annu Meet, pp 15–17

Hatton TT, Cubbedge RH, Risse LA, Hale PW (1982) Phytotoxicity of gamma irradiation on Florida grapefruit. Proc Fla State Hortic Soc 95: 232–234

Hatton TT, Cubbedge RH, Risse LA, Hale PW, Spalding DH, von Windeguth D, Chew V (1984) Phytotoxic responses of Florida grapefruit to low-dose irradiation. J Am Soc Hortic Sci 109: 607–610

Hayashi T, Kawashima K (1983) Impedance measurement of irradiated potatoes. J Food Sci Technol, Japan 30: 51–54

Haynes RH (1966) The interpretation of microbial inactivation and recovery phenomena. Radiat Res Suppl 6: 1–29

Hearn TL, Sgoutas SA, Sgoutas DS, Hearn JA (1987) Stability of polyunsaturated fatty acids after microwave cooking of fish. J Food Sci 52: 1430–1431

Heath RL, Tappel AL (1976) A new sensitive assay for the measurements of hydroperoxides. Anal Biochem 76: 184–191

Heaton EK, Shewfelt AL (1976) Pecan quality. Effect of light exposure on kernel color and flavor. Lebensm Wiss Technol 9: 201–206

Heins HG (1977) Irradiated onions, fresh broilers and mangoes in practice, present status in the Netherlands. In: Biological science. Proc Worksh on the Use of ionizing radiation in agriculture. Wageningen, EUR 5815 EN, pp 171–177

Heins HG, Langerak DI (1980) Quality evaluation of heat treated and/or irradiated papayas and strawberries. An airborne trial shipment from South Africa. Assoc Euratom-ITAL, Prelim Tech Rep 91, Wageningen

Heist J, Cremer ML (1990) Sensory quality and energy use for baking of molasses cookies prepared with bleached and unbleached flour and baked in infrared, forced air convection, and conventional deck ovens. J Food Sci 55: 1095–1101

Herschel W (1800) Experiments on the refrangibility of invisible rays of the sun. Philos Trans R Soc Lond 90: 255–253, 284–292, 293–326

Hicks CL, Abdullah C (1987) Photooxidation of cinnamon oil. J Food Sci 52: 1041–1046

Hidaka A, Suzuki H, Hayakawa S, Okazaki E, Wada S (1989) Conversion of provitamin D_3 to vitamin D_3 by monochromatic ultraviolet rays in fish dark muscle mince. J Food Sci 54: 1070–1073

Hobbs BC (1976) Microbiological hazards of international trade. In: Skinner FA, Carr JC (eds) Microbiology in agriculture fisheries and food. Academic Press, London, pp 161–180

Hoffman CJ, Zabik ME (1985) Effects of microwave cooking/reheating on nutrients and food systems: review of recent studies. J Am Diet Assoc 85: 922–926

Holm NW, Berry RJ (1970) Manual on radiation dosimetry. Marcel Dekker, New York

Hooker WJ, Duncan DT (1959) Storage rot susceptibility of potato tubers exposed to ionizing irradiation. Am Potato J 36: 162–172

Hoskin JC, Dimik PS (1979) Evaluation of fluorescent light on flavor and riboflavin content of milk held in gallon returnable containers. J Food Prot 42: 105–109

Huang HF, Yates RA (1980) Radio-frequency drying of fungal material and resultant textured product. J Microwave Power 15: 15–18

Huang YW, Toledo R (1982) Effect of high doses of high and low intensity UV irradiation on surface microbiological counts and storage-life of fish. J Food Sci 47: 1667–1669

Huet R (1974) Retention des aromes dans les poudres de fruits tropicaux obtenues dans un four a micro-ondes sous vide. Fruits 29: 399–405

Huhtannen CN, Jenkins RK, Thayer DW (1989) Gamma radiation sensitivity of *Listeria monocytogenes*. J Food Prot 52: 610–613

Hulls PJ (1982) Development of the industrial use of dielectric heating in the United Kingdom. J Microwave Power 17: 30–34

Hunter JE, Buddenhagen IW, Kojima ES (1969) Efficacy of fungicides, hot water and gamma-irradiation for control of postharvest fruit rots of papaya. Plant Dis Rep 53: 279–284

Huxsoll CC, Dietrich WC, Morgan AI Jr (1970) Comparison of microwave with stream or water blanching of corn-on-the-cob. I. Characteristics of equipment and heat penetration. Food Technol 24: 290–293

Huyzers CJ, Basson R (1985) The radurisation of bananas, SAFFOST '85 Congr, Pretoria, vol 2, pp 473–475

IAEA-TECDOC-587 (1991) Analytical detection methods for irradiated foods. Vienna

Ichikawa S (1981) Responses to ionizing radiation. In: Lang OL, Nobel PS, Osmond CB, Ziegler H (eds) Physiological plant ecology. I. Responses to the physical environment. Springer, Berlin Heidelberg New York, pp 199–228

Idziak ES, Incze K (1968) Radiation treatment of foods. I. Radurization of fresh eviscerated poultry. Appl Microbiol 16: 1061–1066

Idziak ES, Whitaker RS (1969) Alteration of microbial population of poultry after radurization and storage treatment. Bacteriol Proc 13

Ingram M, Farkas J (1977) Microbiology of foods pasteurized by ionizing radiation. Acta Aliment 6: 123–185

Izat AL, Gardner FA, Denton JH, Golan FA (1988) Incidence and level of *Campylobacter jejuni* in broiler processing. Poult Sci 67: 1568–1572

Jacobs CJ, Brodrick HT, Swarts HD, Mulder NJ (1973) Control of postharvest decay of mango fruit in South Africa. Plant Dis Rep 57: 173–176

Janky DM, Oblinger JL (1976) Microwave versus water-bath pre-cooking of turkey rolls. Poult Sci 55: 1549–1567

Jaynes HO (1975) Microwave pasteurization of milk. J Milk Food Technol 38: 386–387

Jemmali M, Guilbot A (1970) Influence of gamma irradiation on the tendency of *A. flavus* spores to produce toxins during culture. Food Irradiat 10: 15–19

Jiravatana V, Cuevas-Ruiz J, Graham HD (1970) Extension of storage life of papayas grown in Puerto Rico by gamma radiation treatments. J Agric Univ Puerto Rico 54: 314–319

Johansen I (1975) The radiobiology of strand breakage. In: Hanawalt PC, Setlow RB (eds) Molecular mechanisms for repair of DNA. Plenum Press, New York

Johnson GI, Boag TS, Cook AW, Izard M, Panitz M, Sangchote S (1990) Interaction of post-harvest disease control treatments and gamma irradiation on mangoes. Ann Appl Biol 116: 245–257

Johnson JL, Doyle MP, Cassens RG (1990) *Listeria monocytogenes* and other *Listeria* spp. in meat and meat products. A review. J Food Prot 53: 81–91

Jona R, Fronda A (1990) Rapid differentiation between gamma-irradiated and non-irradiated potato tubers. Radiat Phys Chem 35: 317–320

References

225

Josephson ES, Peterson MS (eds) (1983) Preservation of foods by ionizing radiation, vols 1–3. CRC Press, Boca Raton

Kader AA (1986) Potential applications of ionizing radiation in postharvest handling of fresh fruits and vegetables. Food Technol 40: 117–121

Kadir RA, Bargman TJ, Rupnow JH (1990) Effect of infrared heat processing on rehydration rate and cooking of *Phaseolus vulgaris* (var. Pinto). J Food Sci 55: 1472–1473

Kaess G, Weidemann JF (1973) Effects of ultraviolet irradiation on the growth of microorganisms on chilled beef slices. J Food Technol 8: 59–69

Kahan RS, Barkai-Golan R (1968) Combined action of sodium orthophenylphenate and gamma radiation on the in vitro development of fungi pathogenic to citrus fruits. Phytopathology 58: 700–701

Kahan RS, Monselise SP (1965) Extension of storage life of citrus fruits by irradiation. Food Technol 19: 122–124

Kahan RS, Padova R (1968) Irradiated citrus fruits and their external appearance during prolonged storage. IA-1160, Israel Atomic Energy Commission, 24

Kahan RS, Monselise SP, Riov J, van Kooy J, Chadwick K (1968a) Comparison of the effect of radiation of various penetrating powers on the damage to citrus fruit peel. Radiat Bot 8: 415–423

Kahan RS, Shiffman-Nadel M, Temkin-Gorodeiski N, Eisenberg E, Zauberman G, Aharoni Y (1968b) Effect of radiation on the ripening of bananas and avocado pears. In: Preservation of fruit and vegetables by radiation. International Atomic Energy Agency, Vienna, pp 3–11

Kaindl K (1966) International project on the irradiation of fruit and fruit juices. IAEA/SM-73. Proc Symp Food irradiation, International Atomic Energy Agency, Vienna, pp 701–713

Kamali AR, Maxie EC, Rae HL (1972) Effect of gamma irradiation on Fuerte avocado fruits. Hortic Sci 7: 125–126

Kampelmacher EH (1977) Irradiation of food – a new technology for preserving and assuring the hygienic quality of foods. Fleischwirtschaft 64: 322–327

Kampelmacher EH (1983) Irradiation for control of *Salmonella* and other pathogens in poultry and fresh meats. Food Technol 37: 117–119

Karam LR, Simic MG (1988) Detecting irradiated foods: use of hydroxyl radical biomarkers. Anal Chem 60: 1117A–1119A

Kase KR, Bjarngard BE, Attix FH (1987) The dosimetry of ionizing radiation. Academic Press, New York

Katusin-Razem B, Mihaljevic B, Razem D (1990) Lipid test. Nature (Lond) 345: 584

Kawamura Y, Uchiyama S, Saito Y (1989a) A half-embryo test for identification of gamma-irradiated grapefruit. J Food Sci 54: 379–382

Kawamura K, Uchiyama S, Saito Y (1989b) Improvement of the half-embryo test for detection of gamma-irradiated grapefruit and its application to irradiated oranges and lemons. J Food Sci 54: 1501–1504

Kearsley MW, Rodriguez N (1981) The stability and use of natural colours in foods: anthocyanin, β-carotene and riboflavin. J Food Technol 16: 421–431

Kenyon EM, Westcott DE, la Casse P, Gould JW (1971) A system for countinous thermal processing of food pouches using microwave energy. J Food Sci 36: 289–293

Khalil H, Villota R (1989) The effect of microwave sublethal heating on the ribonucleic acids of *Staphylococcus aureus*. J Food Prot 52: 544–548

Khan MA, Novak AF, Rao RM (1976) Reduction of polychlorinated biphenyls in shrimp by physical and chemical methods. J Food Sci 41: 262–267

226 References

Khan MA, Vandermey PA (1985) Quality assessment of ground beef patties after infrared heat processing in a conveyorized tube broiler for food service use. J Food Sci 50: 707–709

King AJ, Bosch N (1990) Effect of NaCl and KCl on rancidity of dark turkey meat heated by microwave. J Food Sci 55: 1549–1551

Kirkpatrick RL (1975) Infrared reduction for control of lesser grain borers and rice weevils in bulk wheat. J Kans Entomol Soc 48: 100–107

Kirkpatrick RL, Yancey DL, Marzke FO (1970) Effectiveness of green and ultraviolet light in attracting stored-product insects to traps. J Econ Entomol 63: 1853–1857

Kirkpatrick RL, Bower JH, Tilton EW (1973) Gamma, infrared and microwave radiation combinations for control of *Rhyzopertha dominica* in wheat. J Stored Prod Res 9: 19–24

Kiss I, Farkas J (1982) Radurization of whole eviscerated chicken carcass. Acta Aliment Acad Sci Hung 1: 73–86

Klinger I, Welgreen H, Basker D (1980) Microbiological contamination of fresh broiler chicken meat in a market. Refu Vet 37: 97–101

Klinger I, Fuchs V, Basker D, Malenky E, Barkat G, Egoz N (1981) Microbiological quality of industrially processed frozen broiler chickens in Israel. Refu Vet 38: 136–148

Klinger I, Fuchs V, Basker D, Juven BJ, Lapidot M, Eisenberg E (1986) Irradiation of broiler chicken meat. Isr J Vet Med 42: 181–193

Knutson KM, Marth EH, Wagner MK (1988) Use of microwave ovens to pasteurize milk. J Food Prot 51: 715–719

Korycka-Dahl M, Richardson T (1978) Photogeneration of superoxide anion in serum of bovine milk and in model systems containing riboflavin and amino acids. J Dairy Sci 400–407

Kouzeh KM, Zuilichem DJ, Roozen JP, Pilnik W (1984) Infrared processing of maize germ. Lebensm Wiss Technol 17: 237–239

Krug W (1985) Social costs of *Salmonella* infection in human and domestic animals. In: Snoyenbos GH (ed) International Symposium on *Salmonella*. New Orleans, pp 304–305

Krug W, Rehm N (1983) Nutzen-Kosten Analyse der Salmonellosebekämpfung. Schriftreihe des Bundesministers für Jugend, Familie und Gesundheit, Band 131. W Kohlhamer, Stuttgart

Kuc JA (1976) Phytoalexins. In: Heitfuss R, Williams PH (eds) Encyclopedia of plant physiology. New Series 4. Springer, Berlin Heidelberg New York, pp 632–652

Kuhn GD, Merkley MS, Dennison RA (1968) Irradiation inactivation of brown rot infection on peaches. Food Technol 22: 903–904

Kwiatek K, Wojton B, Stern NJ (1990) Prevalence and distribution of *Campylobacter* spp. on poultry and selected red meat carcasses in Poland. J Food Prot 53: 127–130

Lahellec C, Colin P, Bennejean G, Paquin J, Guillerm A, Debois C (1986) Influence of resident *Salmonella* on contamination of broiler flocks. Poult Sci 65: 2034–2039

Lakritz L, Maerker G (1989) Effect of ionizing radiation on cholesterol in aqueous dispersion. J Food Sci 54: 1569–1572

Lambert JD, Maxcy RB (1984) Effect of gamma radiation of *Campylobacter jejuni*. J Food Sci 49: 665–667

Lambert JH (1760) Photometria Siva de Mensura et Grandibus Luminis, Colorum et Umbrae. Augsburg

References

Lammerding AM, Gracia MM, Mann ED, Robinson Y, Dorward WJ, Troscott RB, Tittiger F (1988) Prevalence of *Salmonella* and thermophilic *Campylobacter* in fresh beef veal and poultry in Canada. J Food Prot 51: 47–52

Langerak DI (1982) Combined heat and irradiation treatments to control mould contamination in fruit and vegetables. Tech and Prelim Res Pap 93, Res Inst ITAL, Wageningen, Netherlands

Langerak DI (1984) The effect of combined treatment on the inactivation of moulds in fruits and vegetables. Food Irradiat Newslett 8: 16–17

Larrigaudière C, Baccaunaud M, Raymond J, Pech JC (1987) Effects of ionization on the physiology, fungal contamination and quality of cold-stored raspberries. Fruits 42: 597–602

Laster H (1954) The "oxygen-effect" in ionizing irradiation. Nature (Lond) 174: 753

Latarjet R (1954) Spontaneous and induced cell restorations after treatments with ionizing and non-ionizing irradiations. Acta Radiol 41: 84–100

Latarjet R, Ephrussi B (1949) Courbes de survie de levures haploides et diploides soumises aux rayons X. CR Acad Sci (Paris) 229: 306–308

Latimer JM, Matsen JM (1977) Microwave oven irradiation as a method for bacterial decontamination in a clinical microbiology laboratory. J Clin Microbiol 6: 340–345

Laubli MW, Bruttel PA, Schalch E (1986) Determination of the oxidative stability of fats and oils: comparison between the active oxygen method and the rancimat method. J Am Oil Chem Soc 63: 792–795

Lea JS, Dodd NJF, Swallow AJ (1988) A method for testing for irradiation of poultry. Int J Food Sci Technol 23: 625–633

Lee EC, Min DB (1988) Quenching mechanism of β-carotene on the chlorophyll sensitized photooxidation of soybean oil. J Food Sci 53: 1894–1895

Lee MK, Kim JB, Byun MW, Kwon JH, Cho HO (1985) Cooking properties of gamma irradiated chicken. J Korean Soc Food Nutr 14: 151–156

Leung HK, Steinberg MF, Wei LS, Nelson AI (1976) Water binding of macromolecules determined by pulsed NMR. J Food Sci 41: 297–300

Li CF, Bradley RL Jr (1969) Degradation of chlorinated hydrocarbon pesticides in milk and butter oil by ultraviolet energy. J Dairy Sci 52: 27–30

Licciardello JJ, Nickerson JR, Golblith SA (1970) Inactivation of *Salmonella* in poultry with gamma radiation. Poult Sci 49: 663–675

Lillard HS (1988) Comparison of sampling methods and implications for bacterial decontamination of poultry carcasses by rinsing. J Food Prot 51: 405–408

Lillard HS (1989) Factors affecting the persistence of *Salmonella* during the processing of poultry. J Food Prot 52: 829–832

Lillard HS (1990) The impact of commercial processing procedures on the bacterial contamination of broiler carcasses. J Food Prot 53: 202–204

Lin CC, Li CF (1971) Microwave sterilization of oranges in glass-pack. J Microwave Power 6: 45–47

Lin W, Sawyer C (1988) Bacterial survival and thermal responses of beef loaf after microwave processing. J Microwave Power Electromag Energy 23: 183–194

Lin YE, Anantheswaran RC (1988) Studies on popping of popcorn in a microwave oven. J Food Sci 53: 1746–1749

Lindsay RE, Krissinger WA, Fields BF (1986) Microwave vs. conventional oven cooking of chicken: relationship of internal temperature to surface contamination by *Salmonella typhimurium*. J Am Diet Assoc 86: 373–374

Little AC (1976) Physical measurements as predictors of visual appearance. Food Technol 30(10): 74–82

228   References

Loaharanu P (1971) Recent research on the influence of irradiation on certain tropical fruits in Thailand. In: Disinfestation of fruit by irradiation. International Atomic Energy Agency, Vienna, pp 113–124

Lorenz K (1975) Irradiation of cereal grain products. CRC Crit Rev Food Sci Nutr 6: 317–382

Lorenz K, Charman E, Dilsaver W (1973) Baking with microwave energy. Food Technol 27: 28–36

Loy HW, Haggerty JF, Combs EL (1951) Light destruction of riboflavin in bakery products. Food Res 16: 360–364

Lu JY, Stevens C, Yakubu P, Loretan PA, Eakin D (1988) Gamma, electron beam and ultraviolet radiation on control of storage rots and quality of walla onions. J Food Prot 12: 53–62

Lu QD (1988) Preliminary investigation of irradiation with ^{60}Cobalt for preservation of *Pleurotus ostreatus*. Zhongguo Shiyongjun (edible fungi of China) 4, pp 13–14 (Hortic Abstr 59 No 4960)

Luby JM, Gray JI, Harte BR, Ryan TC (1986) Photooxidation of cholesterol in butter. J Food Sci 51: 904–908

Lunt RE (1985) The radurisation of sub-tropical fruit. SAFFOST '85 Congr, Pretoria, vol 2, pp 468–472

Madeira K, Penfield MP (1985) Turbot fillet sections cooked by microwave and conventional heating methods: objective and sensory evaluation. J Food Sci 50: 172–177

Mack TE, Heldman DR, Singh RP (1976) Kinetics of oxygen uptake in liquid foods. J Food Sci 41: 309–312

Mahmood T (1972) Effect of ionizing radiation on *Geotrichum candidum* to prolong the storage life of lemon. Plant Dis Rep 56: 582–585

Martel R, N'Soukpoe-Kossi CN, Paquin P, Leblanc RM (1987) Photoacoustic analysis of some milk products in ultraviolet and visible light. J Dairy Sci 70: 1822–1827

Martin DJ, Tsen CD (1981) Baking high-ratio white layer cakes with microwave energy. J Food Sci 46: 1507–1513

Massey LM Jr, Robinson WB, Spaid JF, Splittstoesser DF, van Buren JP, Kertesz ZI (1965) Effect of gamma radiation upon cherries. J Food Sci 30: 759–765

Matches JR, Liston JJ (1968) Growth of *Salmonella* on irradiated and non-irradiated seafoods. J Food Sci 33: 406–410

Mathee FN, Marais PG (1963) Preservation of food by means of gamma rays. Food Irradiat 4: A10–A11

Matheson MS, Dorfman LM (1965) Pulse radiolysis, MIT Press, Cambridge, MA

Mathews-Roth HM, Wilson T, Fujimori E, Krinski N (1974) Carotenoid chromophore length and protection against photosensitization. Photochem Photobiol 19: 217–223

Mathur PB (1963) Low-dose irradiation of fresh fruits. Food Irradiat 4: A26–A28

Mathur PB (1968) Application of atomic energy in the preservation of fresh fruits and vegetables. Indian Food Packer 22: 4–5

Mathur PB, Lewis NF (1961) Storage behaviour of gamma irradiated mangoes. Int J Appl Radiat Isot 11: 43–45

Matsuyama A (1978) Variation of combined heat-irradiation effects on cell in-activation of different types of vegetative bacteria. In: Food preservation by irradiation, vol 1. International Atomic Energy Agency, Vienna, pp 251–262

Matsuyama A, Umeda K (1983) Sprout inhibition in tubers and bulbs. In: Josephson ES, Peterson MS (eds) Preservation of food by ionizing radiation, vol III. CRC Press, Boca Raton, pp 159–213

References

Maurer RL, Tremblay MR, Chadwick EA (1972) Microwave oven occupies 75% less floor space-dries 2000lb pasta/h, cuts process time 95%. Food Process 33: 18–20

Maxcy RB (1983) Significance of the residual organisms in foods after substerilization doses of gamma irradiation: a review. J Food Safety 5: 203–211

Maxie EC, Abdel-Kader A (1966) Food irradiation – physiology of fruits as related to the feasibility of the technology. In: Chichester CO, Mrak EM, Stewart GF (eds) Advances in food research, vol 15. Academic Press, New York, pp 105–145

Maxie EC, Nelson KE, Johnson CF (1964a) Effect of gamma irradiation on table grapes. Proc Am Soc Hortic Sci 84: 263–268

Maxie EC, Sommer NF, Rae HL (1964b) Effect of gamma irradiation on Shasta strawberries under marketing conditions. Food Irradiat 2: 50–54

Maxie EC, Eaks IL, Sommer NF, Rae HL, El-Batal S (1965) Effect of gamma radiation on rate of ethylene and carbon dioxide evolution by lemon fruit. Plant Physiol 40: 407–409

Maxie EC, Johnson CF, Boyd C (1966) Effect of gamma irradiation on ripening and quality of nectarines and peaches. Proc Am Soc Hortic Sci 89: 91–99

Maxie EC, Amezquita R, Hassan BM, Johnson CF (1969) Effect of gamma irradiation on the ripening of banana fruits. Proc Am Soc Hortic Sci 92: 235–254

Maxie EC, Sommer NF, Mitchell FG (1971) Infeasibility of irradiating fresh fruits and vegetables. Hortic Sci 6: 202–204

McGee H (1990) Recipe for safer sauces. Nature (Lond) 347: 717

McLaughlin WL (1982) Trends in radiation dosimetry. Int J Appl Radiat Isot 33: 953–1310

McLaughlin WL, Boyd AW, Chadwick KN, McDonald JC, Miller A (1989) Dosimetry for radiation processing. Taylor and Francis, London.

McWeeny DJ, Scotter SL, Wood R, Dennis MJ (1991) Potential new methods of detection of irradiated food. EUR 13331, BCR Information Seri, Office for Official Publications of the European Communities, Luxembourg, pp 100–101

Meier W, Biedermann M (1990) Detection of irradiated fatty foods by coupled LC-GC. Mitt Geb Lebensmittelunters Hyg 81: 39–50

Meier W, Burgin R, Frohlich D (1989) Analysis of o-tyrosine as a method for identification of irradiated fresh meat (chicken). Mitt Geb Lebensmittelunters Hyg 80: 22–29

Meier W, Burgin R, Frohlich D (1990) Analysis of o-tyrosine as a method for the identification of irradiated chicken and the comparison with other methods (analysis of volatiles and ESR-spectroscopy). Radiat Phys Chem 35: 332–336

Mercier RG, MacQueen KF (1965) Gamma irradiation to extend postharvest life of fruits and vegetables. Rep Horticultural Experiment Station and Product Laboratory, Vineland Station, Ontario, pp 52–72

Meredith DS (1960) Studies on the *Gloeosporium musarum* Oke and Masse causing storage rots in Jamaican bananas. I. Anthracnose and its chemical control. Ann Appl Biol 48: 279–290

Metaxas AC (1976) Rapid heating of liquid foodstuffs at 896 MHz. J Microwave Power 11: 105–109

Methta RS, Bassette R (1979) Volatile compounds in UHT-sterilized milk during fluorescent light exposure and storage in the dark. J Food Prot 42: 256–258

Michener HD, Elliot RP (1964) Minimum growth temperatures for food poisoning, fecal indicator, and psychrophilic microorganisms. In: Chichester CO, Mrak EM, Stewart GF (eds) Advances in food research, vol 13. Academic Press, New York, pp 349–396

230 References

Miller BJ, Billedeau SM, Miller W (1989) Formation of N-nitrosamines in microwaved versus skillet-fried. Food Chem Toxicol 27: 295–299

Mohr H, Drumm-Herrel H (1981) Interaction between blue/UV light and light operating through the phytochrome in higher plants. In: Smith H (ed) Plants and the daylight spectrum. Academic Press, London pp 423–442

Mohr H, Drumm-Herrel H (1983) Coaction between phytochrome and blue/UV light in anthocyanine synthesis in seedlings. Physiol Plant 58: 408–414

Morehouse KM, Ku Y (1990) A gas chromatographic method for the identification of gamma-irradiated frog legs. Radiat Phys Chem 35: 337–341

Morehouse KM, Ku Y, Albrecht HL, Yang GC (1991) Gas chromatographic and electron spin resonance investigations of gamma-irradiated frog legs. Radiat Phys Chem 38: 61–68

Morris LL, Herner R, Abdel-Kader AS (1964) Studies of gamma irradiation effects on storage life of cucumbers. In: Radiation technology in conjunction with postharvest procedures as a means of extending the shelf life of fruits and vegetables. Rep 1963 to 1964. Div Isotop Dev, US Atomic Energy Comission, Washington DC

Moshonas MG, Shaw PE (1982) Irradiation and fumigation effects on flavor, aroma and composition of grapefruit products. J Food Sci 47: 958–960

Mossel DAA (1977) Microbiology of foods: occurrence, prevention, and monitoring of hazards and deteriorations. Univ Press, Utrecht

Mossel DAA (1987) Processing for safety of meat and poultry by radicidation: progress, penury, prospects. In: Smulder FM (ed) Elimination of pathogenic organisms from meat and poultry. Elsevier, New York, pp 305–316

Moy JH (1977a) Combined treatment of UV and gamma-radiation of papaya for decay control. FAO/IAEA/WHO Int Symp on Food Preserv by Irrad, 21–25 Nov, Wageningen

Moy JH (1977b) Potential of gamma irradiation of fruits: a review. J Food Technol 12: 449–457

Moy JH (1983) Radurization and radicidation: fruits and vegetables. In: Josephson ES, Peterson MS (eds) Preservation of food by ionizing radiation, vol III. CRC Press, Boca Raton, pp 83–108

Moy JH, Chang GKL, Hsia ST (1969) Organoleptic evaluation of gamma irradiated lychee. In: Dosimetry, tolerance and shelf-life extension related to disinfestation of fruits and vegetables by gamma irradiation. Rep 1967 to 1968. Div Isotop Dev, US Atomic Energy Commission, Washington DC

Moy JH, Akamine EK, Wenkam N, Dollar AM, Hanaoka M, Kao HY, Liu WL, Rovettii LM (1973) Tolerance, quality and shelf-life of gamma-irradiated papaya grown in Hawaii, Taiwan and Venezuela. In: Radiation preservation of food, International Atomic Energy Agency, Vienna, pp 375–387

Moy JH, McElhaney T, Matsuzaki C, Piedrahita C (1978) Combined treatment of UV and gamma-radiation of papaya for decay control. In: Proc Int Symp Food Preservation Irradiat vol I. ST1/PUB/470, IAEA, Vienna, pp 361–368

Mudgett RE (1986) Microwave properties and heating characteristics of foods. Food Technol 40: 84–93

Mudgett RE (1989) Microwave food processing. Food Technol 41: 117–126

Mudgett RE, Goldblith SA, Wang DIC, Westphal WB (1977) Prediction of dielectric properties in solid foods of high moisture content at ultra-high and microwave frequency. J Food Proc Preserv 1: 119–122

Mulder RWAW (1982) The use of low temperature and radiation to destroy enterobacteriaceae and salmonellae in broiler carcasses. J Food Technol 17: 461–466

References   231

Mulder RWAW (1989) Current prospects for decontamination of raw and cooked poultry products. Food Aust 41: 1090–1093

Mulder RWAW, Notermans S, Kampelmacher EH (1977) Inactivation of *Salmonella* on chilled and deep frozen broiler carcasses by irradiation. J Appl Bacteriol 42: 179–185

Muller WH (1956) Influence of temperature on growth and sporulation of certain fungi. Bot Gaz 117: 336–343

Mullins WR, Burr HK (1961) Treatment of onions with gamma rays: effects of delay between harvest and irradiation. Food Technol 15: 178–179

Mumtaz A, Farooqui WA, Sattar A, Muhammad A (1970) Inhibition of sproutsing in onions by gamma radiation. Food Irradiat 10: 10–14

Musco DD, Cruess WV (1954) Food rancidity. Studies in deterioration of walnut meats. J Agric Food Chem 2: 520–523

Nair PM, Thomas P, Ussuf KK, Surendranathan KK, Limaye SP, Srirangarajan AN, Padwal Desai SR (1973) Studies of sprout inhibition of onions and potatoes and delayed ripening of bananas and mangoes by gamma irradiation. In: Radiation preservation of food. International Atomic Energy Agency, Vienna, pp 347–366

Nakayama TOM, Allen JM, Cummins S, Wang YYD (1983) Disinfestation of dried foods by focused solar energy. J Food Process Preserv 7: 1–6

Nawar WW (1977) Radiation chemistry of lipids. In: Elias PS, Cohen AJ (eds) Radiation chemistry of major food components. Elsevier, Amsterdam, pp 21–61

Nawar WW, Balboni JJ (1970) Detection of irradiation treatment in foods. J Assoc Off Anal Chem 53: 726–729

Nelson KE, Maxie EC, Eukel W (1959) Some studies on the use of ionizing radiation to control *Botrytis* rot in table grapes and strawberries. Phytopathology 49: 475–480

Nelson SO (1976) Microwave dielectric properties of insect and grain kernels. J Microwave Power 11: 299–305

Nelson SO (1978) Electrical properties of grain and other food materials. J Food Process Preserv 2: 137–141

Nelson SO, Payne JA (1982) Pecan weevil control by dielectric heating. J Microwave Power 17: 51–59

Nolfisinger GW, Vaneauwenberge JE, Anderson RA, Bothost RJ (1980) Preliminary biological evaluation of the effect of microwave heating on high moisture shelled corn. Cereal Chem 57: 373–377

Notermans S, Kampelmacher EH (1974) Attachment of some bacterial strains to the skin of broiler chicken. Br Poult Sci 15: 573–585

Nuttall VW, Lyall LH, MacQueen KF (1961) Some effects of gamma radiation on stored onions. Can J Plant Sci 41: 805–813

Nyambati MGO, Langerak DI (1984) Effects of mild heat, potassium metabisulfite and/or gamma irradiation on storability of lime fruits. Acta Aliment 13: 65–81

Ockermann HW, Cahill VR, Plimpton RF, Parrett NA (1976) Cooking inoculated pork in microwave and conventional ovens. J Food Milk Technol 39: 771–773

Ogbadu G (1980) Influence of gamma irradiation on aflatoxin B_1 production by *Aspergillus flavus* on some Nigeria foodstuffs. Microbios 27: 19–26

Ohlsson T, Bengtsson NE, Risman PO (1974) The frequency and temperature dependence of dielectric food data as determined by a cavity perturbation technique. J Microwave Power 9: 129–135

Ojima T, Hori S, Ueno T, Toratani H, Katayama T, Fujimoto H, Kitoh S, Mori S, Yoshizako F, Nishimura A, Shiomi N, Inoue M (1973) Effects of electron irradiation on the preservation of *Citrus unshiu* Part II: Preservation of fruits wrapped with some kinds of plastic film and irradiation. Food Irradiat Jpn 8: 11–20

Ojima T, Hori S, Ueno T, Toratani H, Katayama T, Fujimoto H, Kitoh S, Yoshizako F, Nishimura A, Shiomi N, Inoue M, Mori S (1974) Effects of electron of electron irradiation of peel by irradiation. Food Irradiat Jpn 9: 1–12

Ojima T, Hori S, Ueno T, Toratani H, Katayama T, Fujimoto H, Kitoh S, Yoshizako F, Nishimura A, Shiomi N, Inoue M, Hasegawa H (1975) Effects of electron irradiation on the preservation of *Citrus unshiu*. Part IV: Storage and browning of peel by irradiation (2) Food Irradiat Jpn 10: 7–13

Ollinger SPA, Matthews ME (1988) Cook/chill foodservice system with a microwave oven: coliforms and aerobic counts from turkey rolls and slices. J Food Prot 51: 84–86

Olsen CM (1965) Microwave inhibit bread molds. Food Eng 37: 51–53

Olsen CM, Drake CL, Banch SL (1966) Some biological effects of microwave energy. J Microwave Power 1: 45–50

O'Mahoney M, Goldstein LR (1978) Sensory techniques for measuring differences in California Naval oranges treated with doses of gamma-radiation below 0.6 kGy. J Food Sci 52: 348–352

O'Meara JP, Shaw TM (1957) Detection of free radicals in irradiated food constituents by electron paramagnetic resonance. Food Technol 11: 132–136

Osborne BG, Fearn T (1986) Near infrared spectroscopy in food analysis. Longman Scientific and Technical, New York

Osterdahl BG, Alriksson E (1990) Volatile nitrosamines in microwave-cooked bacon. Food Addit Contam 7: 51–54

Overview (1989) Food irradiation. Food Technol 43: 75–97

Pace WE, Westphal WB, Goldblith SA (1968) Dielectric properties of commercial cooking oils. J Food Sci 33: 30–42

Padwal-Desai SR, Ghanekar AS, Thomas P, Sreenivasan A (1973) Heat-radiation combination treatment for control of mold infection in harvested fruits and processed cereal foods. Acta Aliment 2: 189–192

Palamides N, Markakis P (1975) Stability of grape anthocyanin in a carbonated beverage. J Food Sci 40: 1047–1049

Palanuk Sl, Warthesen JJ, Smith DE (1988) Effect of agitation, sampling location and protective films on light-induced riboflavin loss in skim milk. J Food Sci 53: 436–438

Palumbo SA, Jenkins RK, Buchanan RL, Thayer DW (1986) Determination of irradiation D-values for *Aeromonas hydrophila*. J Food Prot 49: 189–191

Park YW (1987) Effect of freezing, thawing, drying, and cooking on carotene retention in carrots, broccoli and spinach. J Food Sci 52: 1022–1025

Paster N, Barkai-Golan R (1986) Heat treatment and gamma irradiation effects on ochratoxin production by *Aspergillus ochraceus* sclerotia. Trans Br Mycol Soc 87: 223–228

Paster N, Barkai-Golan R, Padova R (1985) Effects of gamma radiation on ochratoxin production by the fungus *Aspergillus ochraceus*. J Sci Food Agric 36: 445–449

Penfield MP, Costello CA, McNeil MA, Riemann MJ (1989) Effects of fat level and cooking methods on physical and sensory characteristics of restructured beef steaks. J Food Qual 11: 349–356

Peng SK, Taylor B, Tham P, Werthessen NT, Mikkelson B (1976) Effect of autooxidation products from cholesterol on aortic smooth muscle cells. Arch Pathol Lab Med 102: 57–61

References

Pettinati JD (1975) Microwave oven method for rapid determination of moisture in meat. J Assoc Off Anal Chem 58: 1188–1193

Pfeffer PE, Gerasimowicz WV (1989) Nuclear magnetic resonance in agriculture. CRC Press, Boca Raton

Phan PA (1977) Microwave thawing of peaches. A comparative study of various thawing processes. J Microwave Power 12: 261–266

Phillips DJ, Uota M, Monticelli D, Curtis C (1976) Colonization of almond by *Aspergillus flavus*. J Am Soc Hortic Sci 101: 19–23

Pieper H, Stuart JA, Renwick WR (1977) Microwave technique for rapid determination of moisture in cheese. J Assoc Off Anal Chem 60: 1392–1396

Pierce T, Moss E, Sams WM, Akers JH (1986) Hazards of ultraviolet radiation. J Environ Health 49: 76–80

Pini PN, Gilbert RJ (1988) The occurrence in the U.K. of *Listeria* species in raw chickens and soft cheeses. Int J Food Microbiol 6: 317–326

Pluyer HR, Ahmed EM, Wei CI (1987) Destruction of aflatoxins on peanuts by oven- and microwave-roasting. J Food Prot 50: 504–508

Pominski J, Vinnett CH (1989) Production of peanut flour from microwave vacuum-dried peanuts. J Food Sci 54: 187–189

Pordesimo LO, Anantheswaran RC, Fleischmann AM, Lin YE, Hanna MA (1990) Physical properties as indicators of popping characteristics of microwave popcorn. J Food Sci 55: 1352–1355

Powelson RL (1960) Initiation of strawberry fruit rot caused by *Botrytis cinerea*. Phytopathology 50: 491–494

Priyadarshini E, Tulpule PG (1976) Aflatoxin production on irradiated foods. Food Cosmet Toxicol 14: 293–295

Proctor BE, Goldblith SA (1951) Electromagnetic radiation fundamentals and their application in food technology. Adv Food Res 3: 119–208

Proctor VA, Cunningham FE (1983) Composition of broiler meat as influenced by cooking methods and coating. J Food Sci 48: 1696–1699

Prusa KJ, Hughes KV (1986) Cholesterol and selected attributes of pork tenderloin steaks heated by conventional, convection, and microwave ovens to two internal endpoint temperatures. J Food Sci 51: 1139–1140

Prusky D, Keen NT, Sims JJ, Midland SL (1982) Possible involvement of an antifungal diene in the latency of *Colletotrichum gloeosporioides* on unripe avocado fruits. Phytopathology 72: 1578–1582

Prusky D, Keen NT, Eaks I (1983) Further evidence for the involvement of a preformed antifungal compound in the latency of *Collectotricum gloeosporioides* on unripe avocado fruits. Physiol Plant Pathol 22: 189–198

Pryor WA (1976) The role of free radical reactions in biological systems. In: Pryor WA (ed) Free radicals in biology, vol 1. Academic Press, New York, pp 1–50

Pryor WA, Castle L (1984) Chemical methods for the detection of lipid hydroperoxides. Methods Enzymol 105: 293–299

Purcell AE, Walter WM Jr (1988) Comparison of carbohydrate components in sweet potatoes baked by convection heating and microwave heating. J Agric Food Chem 36: 360–362

Quast DG, Karel M (1972) Effects of environmental factors on the oxidation of potato chips. J Food Sci 37: 584–588

Raffi JJ, Agnel JPL (1983) Influence of the physical structure of irradiated starches on their electron spin resonance spectra kinetics. J Phys Chem 87: 2369–2373

Raffi JJ, Belliardo (1991) Potential new methods of detection of irradiated food. EUR 13331, BCR Information Ser, Office for Official Publications of the European Communities, Luxembourg

234 References

Raffi JJ, Agnel JPL, Buscarlet LA, Martin CC (1988) Electron spin resonance identification of irradiated strawberries. J Chem Soc, Faraday Trans 1, 84: 3359–3362

Raffi JJ, Evans JC, Agnel JP, Rowlands CC, Legards G (1989) ESR analysis of irradiated frog's legs and fishes. Int J Appl Radiat Isotop 40: 1215–1218

Raica N Jr, Scott J, Nielson W (1972) The nutritional quality of irradiated foods. Radiat Res Rev 3: 447–454

Rao VS, Vakil UK (1983) Effects of gamma irradiation on flatulence-causing oligosaccharides in green gram (*Phaseolus areus*). J Food Sci 48: 1791–1795

Raper KB, Fennell DI (1965) The genus *Aspergillus*. Williams and Wilkins, Baltimore

Rashid A, Farooqui WA (1984) Studies on the effect of gamma irradiation and maleic hydrazide on the shelf life of mangoes. J Agric Res Pak 22: 151–158

Ravetto D, Morris LL, Johnson CF, Maxie EC (1967) Effect of gamma irradiation on postharvest behavior of the cantaloupe. In: Radiation technology in conjunction with postharvest procedures as a means of extending the shelf life of fruits and vegetables. Rep 1966 to 1967, Div Isot Dev, US Atomic Energy Commission, Washington DC

Ray EE, Berry BW, Thomas JD (1985) Influence of hot-boning, cooking and method of reheating on product attributes of lamb roast. J Food Prot 48: 412–415

Reagon JO, Smith GC, Carpenter ZL (1973) Use of ultraviolet light for extending the retail caselife of beef. J Food Sci 38: 929–931

Reckhagel RO, Glende EA Jr (1984) Spectrophotometric detection of lipid conjugated dienes. Methods Enzymol 105: 331–337

Richardson LR (1955) A long range investigation of the nutritional properties of irradiated food. Progr Rep III, Sept 1 1954–July 1 1955. Texas Agriculture Experiment Station, College Station, Texas. Contract DA49-007-MD-582, Defence Documentation Center, Alexandria, Virginia

Riov J (1971) 6,7-Dimethoxycoumarin in the peel of gamma irradiated grapefruit. Phytochemistry 10: 1923

Riov J (1975) Histochemical evidence for the relationship between peel damage and the accumulation of phenolic compounds is gamma-irradiated citrus fruit. Radiat Bot 15: 257–260

Riov J, Goren R (1970) Effects of gamma radiation and ethylene on protein synthesis in peel of mature grapefruit. Radiat Bot 10: 155–160

Riov J, Monselise SP, Kahan RS (1968) Effects of gamma radiation on phenylalanine ammonia lyase activity and accumulation of phenolic compounds in citrus fruit peel. Radiat Bot 8: 463–466

Riov J, Goren R, Monselise SP, Kahan RS (1971) Effect of gamma irradiation on the synthesis of scopoletin and scopolin in grapefruit peel in relation to radiation damage. Radiat Res 45: 326–334

Riov J, Monselise SP, Goren R, Kahan RS (1972) Stimulation of phenolic biosynthesis in citrus fruit peel by gamma irradiation. Radiat Res Rev 3: 417–427

Risman PO, Bengtsson NE (1971) Dielectric properties of foods at 3 GHz as determined by a cavity perturbation technique. J Microwave Power 6: 101–123

Ritter JW (1801) Physisch-chemische Abhandlungen, vol 2. Leipzig

Roberts D (1990) Sources of infection: food. Lancet 336: 859–861

Roberts RL (1977) Effect of microwave treatment of pre-soaked paddy, brown and white rice. J Food Sci 42: 804–806

Roberts T, Pinner R (1990) Foodborne listeriosis. In: Miller AJ, Smith JL, Somkuti GA (eds) Publ Soc for Industrial Microbiology, cited by FAO/WHO Newslett 26

References   235

Rogachev VI (1966) Use of ionizing radiation to prolong the storage life of fruit and berries (review of work in the USSR). In: Application of food irradiation in developing countries. International Atomic Energy Agency, Vienna, pp 123–142

Rooney ML (1981) Oxygen scavenging from air in package headspace by singlet oxygen reactions in polymer media. J Food Sci 47: 291–294

Rooney ML (1982) Oxygen scavenging: a novel use of rubber photo-oxidation. Chem Ind: 197–198

Rooney ML (1983) Photosensitive oxygen scavenger films: an alternative to vacuum packaging. CSIRO Food Res Q 43: 9–11

Rooney ML, Holland RV, Shorter AJ (1981) Photochemical removal of headspace oxygen by a singlet oxygen reaction. J Sci Food Agric 32: 265–272

Rosenberg U, Bohl W (1987a) Microwave thawing, drying and baking in the food industry. Food Technol 41: 86–91

Rosenberg U, Bohl W (1987b) Microwave pasteurization, sterilization, blanching and pest control in the food industry. Food Technol 41: 92–99

Rosenthal I (1985) Photooxidation of foods. In: Frimer AA (ed) Singlet O_2, vol 4. CRC Press, Boca Raton, pp 145–163

Rosenthal I, Yang GC, Bell SJ, Scher AL (1988) The chemical photosensitizing ability of certified colour additives. Food Addit Contam 5: 563–571

Roy MK (1975) Radiation, heat and chemical combine in the extension of shelf-life of apples infected with blue mold rot (*Penicillium expansum*). Plant Dis Rep 59: 61–64

Roy MK, Bahl N (1984a) Gamma radiation for preservation of *Agaricus bisporus*. Mushroom J 136: 124–125

Roy MK, Bahl N (1984b) Studies on gamma radiation preservation of *Agaricus bisporus*. Mushroom J 144: 411–414

Roy MK, Mukewar P (1973) Combined gamma-irradiation and chemical treatment in the control of *Aspergillus niger* van Tieghem and *Fusarium coeruleum* (Lib.) Sacc. In: Radiation preservation of food. International Atomic Energy Agency, Vienna, pp 193–200

Rzepecka-Stuchly MA (1976) Microwave energy in foam-mat dehydration process. J Microwave Power 11: 255–261

Saiz de Bustamante C, Garcia de Matoes A, Hernandes GE, Safont TA (1970) Use of radiation in the preservation of Spanish orange. Rev Agroquim Tecnol Aliment 10: 371–375

Saks Y, Sonego L, Ben-Arie R (1990) Artific light enhances red pigmentation, but not ripening, of harvested "Anna" apples. Hortic Sci 25: 547–549

Salunkhe DK (1961) Gamma radiation effects on fruits and vegetables. Econ Bot 15: 28–56

Salunkhe DK, Wu MT, Jadhav SJ (1972) Effects of light and temperature on the formation of solanine in potato slices. J Food Sci 37: 969–970

Sanders HR (1966) Dielectric thawing of meat and meat products. J Food Technol 1: 183–192

Sanderson DCW, Slater C, Cairns KJ (1989a) Detection of irradiated food. Nature (Lond) 340: 23–24

Sanderson DCW, Slater C, Cairns KJ (1989b) Thermoluminescence of foods: origins and implications for detecting irradiation. Radiat Phys Chem 34: 915–924

Sapers GM, Taffer I, Ross IR (1981) Functional properties of a food colorant prepared from red cabbage. J Food Sci 46: 105–109

Saravacos G, Macris G (1963) Radiation preservation of grapes and some other Greek fruit. Food Irradiat 4: A19–A21

236   References

Saravacos GD, Hatzipetrou LP, Georgiadou E (1962) Lethal doses of gamma radiation of some fruit spoilage microorganisms. Food Irradiat 3: A6–A9

Sattar A, deMan JM (1975) Photooxidation of milk and milk products: a review. CRC Crit Rev Food Sci Nutr 7: 13–37

Sattar A, deMan JM, Alexander JC (1976) Light-induced oxidation of edible oils and fats. Lebensm Wiss Technol 9: 149–152

Sattar A, deMan JM, Alexander JC (1977) Wavelength effect on light-induced decomposition of vitamin A and β-carotene in solutions and milk fat. J Inst Can Sci Technol Aliment 10: 56–60

Sattar A, Delincee H, Diehl JF (1987) Detection of gamma irradiated pepper and papain by chemiluminescene. Radiat Phys Chem 29:215–218

Satterlee LD, Hansmeyer W (1974) The role of light and surface bacteria in the color stability of prepacked beef. J Food Sci 39: 305–308

Sawyer CA (1985) Post-processing temperature rise in foods: conventional hot air and microwave ovens. J Food Prot 48: 429–434

Sawyer CA, Biglari SD, Thompson SS (1984) Internal end temperature and survival of bacteria on meats with and without chloride wrap during microwave cooking. J Food Sci 49: 972–974

Sawyer RL, Dallyn SL (1961) Effect of irradiation on storage quality of potatoes. Am Potato J 38: 227–235

Schaich KM (1980a) Free radical initiation in proteins and amino acids by ionizing and ultraviolet radiations and lipid oxidation. Part I. Ionizing radiation. CRC Crit Rev Food Sci Nutr 13: 89–130

Schaich KM (1980b) Free radical initiation in proteins and amino acids by ionizing and ultraviolet radiation and lipid oxidation. Part II Ultraviolet radiation and photolysis. CRC Crit Rev Food Sci Nutr 12: 131–159

Schiffmann RF (1986) Food product development for microwave processing. Food Technol 40: 94–98

Schiffmann-Nadel M, Cohen E (1966) Influence of grove temperatures on the effectiveness of heat treatment of *Phytophthora*-infected citrus fruits. Plant Dis Rep 50: 867–868

Schindler AF, Abadie AN, Simpson RE (1980) Enhanced aflatoxin production by *Aspergillus flavus* and *Aspergillus parasiticus* after gamma irradiation of the spore inoculum. J Food Prot 43: 7–9

Schnepf M, Barbeau WE (1989) Survival of *Salmonella typhimurium* in roasting chickens cooked in a microwave, convection microwave and a conventional electric oven. J Food Safety 9: 245–252

Setser CS, Harrison DL, Kropf DH, Dayton AD (1973) Radiant energy induces changes in bovine muscle pigment. J Food Sci 38: 412–417

Shanley RM, Jameson GW (1981) A study of the rapid determination of moisture in cheese by microwave heating. Aust J Dairy Technol 36: 107–109

Shantha T, Murthy VS (1981) Use of sunlight to partially detoxify groundnut (peanut) cake flour and casein contaminated with aflatoxin B_1. J Assoc Off Agric Chem 64: 291–195

Shepered AD (1959) Effect of illumination on color of frozen peas packaged in transparent film. Food Technol 11: 539–540

Sigman M, Burke KI, Swarner OW, Shavlik GW (1989) Effects of microwaving human milk: changes in IgA content and bacterial count. J Am Diet Assoc 89: 690–692

Silliker JH (1982) The *Salmonella* problem: current status and future direction. J Food Prot 45: 661–666

References 237

Silliker JH, Elliott RP, Baird Parker AC, Bryan FL, Christian JHB, Clark DS, Olson JC, Roberts TA (eds) (1980) Microbial ecology of foods. ICMSF, Academic Press, New York

Simard RE, Bourzeis M, Heredia N (1982) Factors influencing color degradation in blueberry juice. Lebensm Wiss Technol 15: 177–180

Singh RP, Heldman DR, Kirk JR (1976) Kinetics of quality degradation: ascorbic acid oxidation in infant formula during storage. J Food Sci 41: 304–308

Skirrow MB (1990) *Campylobacter*. Lancet 336: 921–923

Skou JP (1960) Microbiological studies in connection with irradiation of carrots. Danish Atomic Energy Comm, Riso. Riso Rep 16: 79–83

Skou JP (1964a) *Aureobasidium pullulans* (De By.) Arnaud- a common and very radio-resistant fungus on fresh and vegetables. In: Kinnell PO, Runnstrom-Reio V (eds) Radiation preservation of foodstuffs, 2nd Scand Meet on Food preservation by ionizing radiation, Stockholm, IVA Medd 138, pp 63–70

Skou JP (1964b) Radiation-induced damage to plant tissues as a cause of the intensified attacks by microorganisms following irradiation. In: Kinnell PO, Runnstrom-Reio V (eds) Radiation preservation of foodstuffs, 2nd Scand Meet on Food preservation by ionizing radiation, Stockholm, IVA Meed 138, pp 72–78

Skou JP (1971) Studies on the effects of ionizing radiation for extending the storage lives of onions. Riso Rep 238, 46–49

Skou JP (1977) On the intensified attack by microorganisms following irradiation-induced prout inhibition in vegetables. In: Biological science. Proc Worksh on the Use of ionizing radiation in agriculture. Wageningen, EUR 5815 EN, pp 131–156

Skou JP, Henriksen JB (1964) Increased susceptibility to storage rot in potatoes and carrots after sprout-inhibiting gamma-radiation. In: Kinnell PO, Runnstrom-Reio V (eds) Radiation preservation of foodstuffs. 2nd Scand Meet on Food preservation by ionizing radiation, Stockholm, IVA Medd 138, pp 48–54

Slater TF (1984) Overview of methods for detecting lipid peroxidation. Methods Enzymol 105: 283–293

Sliney DH, Freasier BC (1973) Evaluation of optical radiation hazards. Appl Optics 12: 1–24

Smierzchalska K, Wojniakiewicz E (1986) The effect of ionizing radiation and storage temperature on the postharvest growth and some quality properties of mushrooms (*Agaricus bisporus*). Acta Agrobot 39: 207–219

Smierzchalska K, Perlowska M, Wojniakiewicz E, Habdas H (1988) Application of ionizing radiation for prolonging the shelf-life of certain vegetables. Int Agrophys 4: 339–347

Smierzchalska K, Swiniarski D, Horbowicz M (1989) The preservation of mushrooms by irradiation. Biul Warzywniczy Suppl I: 211–217

Smith FR (1942) The use of ultraviolet rays in the cheese factory and storage room. J Dairy Sci 25: 525–527

Smith LL, Kalig MJ (1975) On the derivation of carcinogenic sterols from cholesterol. Cancer Biochem Biophys 1: 79–84

Smith O (1968) Potatoes: production, storing, processing. Avi, Westport, CT, pp 539–544

Sobels FH (1963) Repair from genetic radiation damage. Macmillan, New York

Solberg M, Franke WC (1971) Photosensitivity of fresh meat color in the visible spectrum. J Food Sci 36: 990–995

Sommer HH (1938) Market milk and related products. Published by the author. Madison, WI, pp 519–521

Sommer NF (1973) The effect of ionizing radiation on fungi. In: Manual on radiation sterilization of medical and biological materials. Techn Rep Ser 149, International Atomic Energy Agency, Vienna, pp 73–79

Sommer NF, Buchanan JR (1978) Mycotoxin production by postharvest pathogens of fruits and vegetables. In: Rosenberg P (ed) Toxins: animal, plant and microbial. Proc 5th Symp Int Soc Toxicology. Pergamon Press, New York, pp 819–828

Sommer NF, Creasy MT (1964) Recovery of *Rhizopus stolonifer* sporangiospores after potentially lethal gamma irradiation. Radiat Res 22: 74–79

Sommer NF, Fortlage RJ (1966) Ionizing radiation for control of postharvest diseases of fruits and vegetables. In: Chichester CO, Mrak EM, Stewart GF (eds) Advances in food research, vol 15. Academic Press, New York, pp 147–193

Sommer NF, Maxie FC (1966) Recent research on the irradiation of fruits and vegetables. In: Proc Int Symp on Food irradiation. International Atomic Energy Agency, Vienna, pp 571–587

Sommer NF, Creasy M, Romani RJ, Maxie EC (1963a) Recovery of gamma irradiated *Rhizopus stolonifer* sporangiospores during autoinhibition of germination. J Cell Comp Physiol 61: 93–98

Sommer NF, Creasy M, Maxie EC, Romani RJ (1963b) Production of pectolytic enzymes by *Rhizopus stolonifer* sporangio spores after "lethal" gamma irradiation. Appl Microbiol 11: 463–466

Sommer NF, Maxie EC, Fortlage RJ (1964a) Quantitative dose response of *Prunus* fruit decay fungi to gamma irradiation. Radiat Bot 4: 309–316

Sommer NF, Maxie FC, Fortlage RJ, Eckert JW (1964b) Sensitivity of citrus fruit decay fungi to gamma irradiation. Radiat Bot 4: 317–322

Sommer NF, Creasy M, Romani RJ, Maxie EC (1964c) An oxygen dependant postirradiation restoration of *Rhizopus stolonifer* sporangiospores. Radiat Res 22: 21–28

Sommer NF, Görtz JH, Maxie EC (1965) Prevention of repair in irradiated *Rhizopus stolonifer* sporangio spores by inhibitors of protein synthesis. Radiat Res 24: 390–397

Sommer NF, Fortlage RJ, Buckley PM, Maxie EC (1967) Radiation-heat synergism for inactivation of market disease fungi of stone fruits. Phytopathology 57: 428–433

Sommer NF, Buckley PM, Fortlage RJ, Coon DA, Maxie EC, Mitchell FG (1968) Heat sensitization for control of grey mold of strawberry fruit by gamma irradiation. Radiat Bot 8: 441–448

Sommer NF, Dupuy P, Rabatu A (1971) Effects of chemical sensitization on repair of potentially lethal irradiation injury in *Rhizopus stolonifer* sporangiospores. Radiat Bot 11: 363–366

Sommer NF, Fortlage RJ, Buckley PM, Maxie EC (1972) Comparative sensitivity to gamma radiation of conidia, mycelia and sclerotia of *Botrytis cinerea*. Radiat Bot 12: 99–103

Sommer NF, Buchanan JR, Fortlage RJ (1974) Production of patulin by *Penicillium expansum*. J Appl Microbiol 28: 589–593

Sommer NF, Buchanan JR, Fortlage RJ (1976) Aflatoxin and sterigmatocystin contamination of pistachio nuts in orchards. Appl Environ Microbiol 32: 64–67

Souda KB, Akyel C, Bilgen E (1989) Freeze dehydration of milk using microwave energy. J Microwave Power Electromag Energy 24: 195–202

References 239

Sowbhagya CM, Bhattacharya KR (1976) Lipid autooxidation in rice. J Food Sci 41: 1018–1023

Sowden RE (1981) Gamma – radiation processing of food. Proc Biochem: 43–49

Spalding DH, Reeder WF (1986a) Influence of hot water and gamma irradiation treatments on bacterial soft rot of tomatos. Proc Fla State Hort Soc 99: 145–148

Spalding DH, Reeder WF (1986b) Decay and acceptability of mangos treated with combinations of hot water, imazalil, and γ-radiation. Plant Dis 70: 1149–1151

Spalding DH, von Windeguth DL (1988) Quality and decay of irradiated mangos. Hortic Sci 23: 187–189

Sparks WC, Iritani WM (1964) The effect of gamma rays from fission products waste on storage losses of Russett Burbank potatoes. Idaho Agric Exp Stn Bull 60

Spencer PL (1945) A heating device using microwaves. US Patent 2,480,679

Spikes JD (1981) Photodegradation of foods and beverages. In: Smith C (ed) Photochemical and photobiological reviews, vol 6. Plenum, New York, pp 39–85

Spikes JD (1989) Photosensitization. In: Smith KC (ed) The science of photobiology, 2nd edn. Plenum, New York, pp 79–110

Spinks JWT and Woods RJ (1976) An introduction to radiation chemistry. Wiley, New York

Spite GT (1984) Microwave-inactivation of bacterial pathogens in various controlled frozen food compositions and in a commercially available frozen food product. J Food Prot 47: 458–462

Stannard CJ, Wood JM (1983) Measurement of residual hydrogen peroxide in preformed food cartons decontaminated with hydrogen peroxide and ultraviolet irradiation. J Food Prot 46: 1074–1077

Stannard CJ, Abbiss JS, Wood JM (1985) Efficiency of treatments involving ultraviolet irradiation for decontaminating packaging board of different surface compositions. J Food Prot 48: 786–789

Stapleton GE, Billen D, Hollaender A (1953) Recovery of X-irradiated bacteria at suboptimal temperatures. J Cell Comp Physiol 41: 345–357

Steinbuch E, Rol W (1986) Developments in the production of fermented and pickled vegetables. In: Thorne S (ed) Developments in food preservation, vol 4. Elsevier, London, pp 251–271

Stephens LC, Chastain MF (1959) Light destruction of riboflavin in partially-baked rolls. Food Technol 13: 527–528

Stermer RA, Lasater-Smith M, Brasington CF (1987) Ultraviolet radiation – an effective bactericide for fresh meat. J Food Prot 50: 108–111

Stern NJ, Kazmi SV (1989) *Campylobacter jejuni.* In: Doyle MP (eds) Foodborne bacterial pathogens. Marcel Dekker New York, pp 71–110

Stevens C, Khan VA, Tang AY, Lu JY (1990) The effect of ultraviolet radiation on mold rots and nutrients of stored sweet potatoes. J Food Prot 53: 223–226

Stevenson MH, Gray R (1989a) An investigation into the effect of sample preparation methods on the resulting ESR signal from irradiated chicken bone. J Sci Food Agric 48: 261–267

Stevenson MH, Gray R (1989b) The effect of irradiation dose, storage time and temperature on the ESR signal in irradiated chicken drumsticks. J Sci Food Agric 48: 269–274

Stevenson MH, Crone AVJ, Hamilton JTG (1990) Irradiation detection. Nature (Lond) 344: 202–203

Stuben M (1973) Studies on the influence of electronic flashes on the mortality and fertility of *Musca domestica* (Dipt., Muscidae). Z Angew Entomol 74: 35–41

Sturgeon R, Matusiewicz H (1989) Present status of microwave sample dissolution and decomposition for elemental analysis. Prog Anal Spectrosc 12: 21–39

Sudarmadji S, Urbain WB (1972) Flavour sensitivity of selected animal protein foods to gamma radiation. J Food Sci 37: 671

Swallow AJ (1977) Chemical effects of irradiation. In: Elias PC, Cohen AJ (eds) Radiation chemistry of major food components. Elsevier, Amsterdam, pp 5–20

Tamamoto Y, Brodsky MH, Baker JC, Ames BN (1987) Detection and characterization of lipid hydroperoxides at picomole levels by high-performance liquid chromatography. Anal Biochem 160: 7–13

Tarkowski JA, Stoffer SCC, Beumer RR, Kampelmacher EH (1984) Low dose gamma radiation of raw meat. I. Bacteriological and sensory effects in artificially contaminated samples. Int J Food Microbiol 1: 13–23

Taub IA, Kaprielian RA, Halliday JW, Walker JE, Angelini P, Merritt C Jr (1979a) Factors affecting radiolytic effects in food. Radiat Phys Chem 14: 639–653

Taub IA, Robbins FM, Simic MG, Walker JE, Wierbick E (1979b) Effect of irradiation on meat proteins. Food Technol 184–193

Temcharoen P, Thilly WG (1982) Removal of aflatoxin B1 toxicity but not mutagenicity by 1 Mrad gamma radiation of peanut meal. J Food Safety 4: 199–205

Templeman GJ (1977) Evaluation of several pulsed NMR techniques for solids-in-fat determination of commercial fats. J Food Sci 42: 432–435

Thayer DW, Lachica RV, Huhtanen CN, Wierbicki E (1986) Use of irradiation to ensure the microbiological safety of processed meats. Food Technol 40: 159–162

Thayer DW, Boyd G, Lipson CA, Hayne WC, Baer SH (1990) Radiation resistance of *Salmonella*. J Ind Microbiol 5: 383–389

Thimijan RW, Heins RD (1983) Photometric, radiometric and quantum light units of measure: a review of procedures for interconversion. Hortic Sci 18: 818–822

Thomas AC (1975) The application of ionizing radiation to the shelf-life extension of mangoes in South Africa. Rep S Afr Atomic Energy Board PEL-244, 15

Thomas AC (1977) Radiation preservation of subtropical fruits. Food Irradiat Newslett 1: 19–21

Thomas MH, Calloway DH (1957) Nutritive value of irradiated turkey. II. Vitamin losses after irradiation and cooking. J Am Diet Assoc 33: 1030–1033

Thomas P (1983) Radiation preservation of foods of plant origin. Part I. Potatoes and other tuber crops. Crit Rev Food Sci Nutr 19: 327–379

Thomas P (1984) Radiation preservation of foods of plant origin. Part II. Onions and other bulb crops. Crit Rev Food Sci Nutr 21: 95–136

Thomas P (1985) Radiation preservation of foods of plant origin. Part III. Tropical fruits: bananas, mangoes, and papayas. Crit Rev Food Sci Nutr 23: 147–205

Thomas P (1986a) Radiation preservation of foods of plant origin. Part IV. Subtropical fruits: citrus, grapes, and avocados. Crit Rev Food Sci Nutr 24: 53–89

Thomas P (1986b) Radiation preservation of foods of plant origin. Part V. Temperate fruits: pome fruits, stone fruits, and berries. Crit Rev Food Sci Nutr 24: 357–400

Thompson JS, Thompson A (1990) In-home pasteurization of raw goat's milk by microwave treatment. J Food Microbiol 10: 59–64

Thornley MJ (1963) Microbiological aspects of the use of irradiation for the elimination of salmonellae from food and feedstuffs. In: Radiation control of salmonellae in food and feed products. Tech Rep Ser 22, International Atomic Energy Agency, Vienna, pp 81–106

References

Tilton EW, Brower JH (1985) Supplemental treatments for increasing the mortality of insects during irradiation of grain. Food Technol 39: 75–79

Tilton EW, Brower JH (1987) Ionizing radiation for insect control in grain and grain products. Cereal Foods World 32: 330–335

Tilton EW, Vardell HH, Jones RD (1983) Infrared heating with vacuum for control of the lesser grain borer (*Rhyzopertha dominica* F.) and rice weevil (*Sitophilus oryzae* L.) infesting wheat. J Georgia Entomol 18: 61–69

Tiwari PN, Burk W (1980) Seed oil determination by pulsed NMR without weighing and drying seeds. J Am Oil Chem Soc 57: 119–121

To E, Mudgett RE, Wang DIC, Goldblith SA, Decareau RV (1974) Dielectric properties of food materials. J Microwave Power 9: 303–306

Tochman LM, Stine CM, Harte BR (1985) Thermal treatment of cottage cheese "in-package" by microwave heating. J Food Prot 48: 932–938

Todd ECD (1989a) Preliminary estimates of cost of foodborne disease in Canada and costs to reduce salmonellosis. J Food Prot 52: 586–594

Todd ECD (1989b) Preliminary estimates of foodborne disease in the United States. J Food Prot 52: 595–601

Toyosaki T, Yamamoto A, Mineshita T (1988) Kinetics of photolysis of milk riboflavin. Milchwissenschaft 43: 143–146

Troup GJ, Pilbrow JR, Hutton DR, Hunter CR, Wilson GL (1989) EPR detection of free radicals in (I) coffee and (II) gamma-ray irradiated foodstuffs. Int J Appl Radiat Isot 40: 1223–1226

Truelsen TA (1963) Radiation pasteurization of fresh fruits and vegetables. Food Technol 17: 100–103

Tsen CC (1980) Microwave energy for bread baking and its effect on the nutritive value of bread: a review. J Food Prot 43: 638–640

Tsen CC, Reddy PRK, Gehrke CW (1977) Effects of conventional baking, microwave baking and steaming on the nutritive value of regular and fortified breads. J Food Sci 42: 402–406

Tuguchi T, Ishizaki S, Tanaka M, Nagashima Y, Amano K (1989) Effect of ultraviolet irradiation on thermal gelation of muscle pastes. J Food Sci 54: 1438–1440

Uchiyama S, Uchiyama M (1979) Free radical production in protein-rich food. J Food Sci 44: 1217–1220

Umeda K, Kawashima K, Sato T, Iba, Y, Nishiura M (1969) Shallow irradiation of *Citrus unshiu* by cathode ray. Part I. Effective pasteurization dose of radiation to *Penicillium digitatum* and the effects of irradiation on the fruit quality. Food Irradiat Jpn 4: 91–100

Unklesbay N, Davis ME, Krause G (1983) Nutrient retention of portioned menu items after infrared and convective heat processing. J Food Sci 48: 869–873

Urbain WM (1978a) Irradiation of meats and poultry. Food Irradiat Inf 8: 14–30

Urbain WM (1978b) Food irradiation. Adv Food Res 24: 155–227

Urbain WM (1983) Radurization and radicidation: meat and poultry. In: Josephson ES, Peterson MS (eds) Preservation of food by ionizing radiation. CRC Press, Boca Raton, pp 2–11

Vajdi M, Merritt C Jr (1985) Identification of adduct radiolysis products from pork fat. J Am Oil Chem Soc 62: 1252–1260

Vajdi M, Nawar WW (1979) Identification of radiolytic compounds from beef. J Am Oil Chem Soc 56: 611–615

Vajdi M, Pereira RR (1973) Comparative effects of ethylene oxide, gamma irradiation and microwave treatments on selected spices. J Food Sci 38: 893–895 van der Linde HJ (1982) Progress in food irradiation: South Africa. Food Irradiat Inf 12: 100–118

Vandermey PA, Khan MA (1987) Thiamin retention and sensory quality of infrared and conventionally broiled beef loin steaks for foodservice use. J Foodservice Syst 4: 143–152

van Kooy JG, Langerak DI (1961) Irradiation of onions. Food Irradiat 2: A6–A7

van Putte K, Vermaas L, van den Enden J, den Hollander C (1975) Relations between pulsed NMR, wide-line NMR and dilatometry. J Am Oil Chem Soc 52: 179–181

Vela GR, Wu JF (1979) Mechanism of lethal action of 2450–MHz radiation on microorganisms. Appl Environ Microbiol 37: 550–553

Vela GR, Wu JF, Smith DW (1976) Effect of 2450 MHz microwave radiation on some soil microorganisms. Soil Sci 121: 44–51

Vogler A, Kunkley H (1982) Photochemistry and beer. J Chem Educ 59: 25–27

von Hippel AR (1954) Dielectric materials and applications. Wiley, New York; Dielectrics and Waves. MIT Press, Cambridge, MA

von Sonntag C (1987) The chemical basis of radiation biology. Taylor and Francis, London

Wadsworth JI, Koltun SP (1986) Physicochemical properties and cooking quality of microwave-dried rice. Cereal Chem 63: 346–348

Waggoner PE (1955) Radiation and resistance of tubers to rot. Am Potato J 32: 448–450

Wahid M and Kovacs E (1980) Shelf-life extension of mushrooms (*Agaricus bisporus*) by gamma irradiation. Acta Aliment 9: 357–366

Waites WM, Harding SE, Fowler DR, Jones SH, Shaw D, Martin M (1988) The destruction of spores of *Bacillus subtilis* by the combined effects of hydrogen peroxide and ultraviolet light. Lett Appl Microbiol 7: 139–140

Wang SL (1987) Microwave oven drying method for total solids determination in tomatoes: collaborative survey. Anal Chem J 70: 758–759

Watanabe H, Aoki S, Sato T (1976) Effects of electron energies on peel browning and shelf-life of *Citrus unshiu*. Food Irradiat Jpn 11: 39–46

Watanabe W, Tape NW (1969) Microwave processing of wieners. 2. Effect on microorganisms. Can Inst Food Sci Technol J 2: 104–107

Watters FL (1976) Microwave radiation for control of *Tribolium confusum* in wheat and flour. J Stored Products Res 12: 19–25

Watts BM (1954) Oxidative rancidity and discoloration in meat. Adv Food Res 5: 1–52

WAVFH (1967) Destination of salmonella contaminated food and feed. Report of a round table Conference, organized by the World Association of Veterinary Food Hygienists, May 3–5 1967

Weast RC (1967) Handbook of chemistry and physics. The Chemical Rubber Co, Cleveland

Weil KO, Moeller TW, Bedford CL, Ubrain WM (1970) Microwave thawing of individually quick frozen red tart cherries prior to pitting. J Microwave Power 5: 188–193

Welch CB, Maxcy RB (1975) Characterization of radiation resistant vegetative bacteria in beef. Appl Microbiol 30: 242–250

Whang K, Peng IC (1988) Photosensitized lipid peroxidation in ground pork and turkey. J Food Sci 53: 1596–1598

Whiting AR (1960) Protection and recovery of the cell from radiation damage. In: Hollander A (ed) Radiation protection and recovery. Pergamon Press, New York, pp 117–156

References   243

WHO (1976) Wholesomeness of irradiated food. Summaries of data considered by the Joint FAO/IAEA/WHO Expert Committee on the Wholesomeness of Irradiated Food, Geneva 31 August–7 September WHO/FOOD ADD/77.45. Geneva, Switzerland

WHO (1979) Report of an FAO/WHO working group on microbiological criteria for foods, Geneva, 20–26 February 1979

WHO (1981a) Report of the WHO/WAVFH round table conference on the present status of the *Salmonella* problem (prevention and control). Bilthoven, The Netherlands, 6–10 October 1980, VPH/81/27

WHO (1981b) Economic aspects of communicable diseases. Report on a WHO working group. EURO reports and studies 68. Regional Office for Europe, WHO, Copenhagen

WHO (1981c) Wholesomeness of irradiated food. Summaries of data considered by the Joint FAO/IAEA/WHO Expert Committee on the wholesomeness of irradiated food, EHE 81.24, Geneva, Switzerland

WHO (1982) Guidelines for organization and management of surveillance of foodborne diseases. VPH/82.39

WHO (1983) Christian, JHB (ed) Microbiological criteria for foods, summary of recommendations of FAO/WHO expert consultations and working groups 1975–1981. VPH/83.54

WHO (1984) Report of the WHO, consultation on the veterinary public health aspects of prevention and control of *Campylobacter* infections, Moscow, 20–22 February 1984, VPH/CDD/FOS/84.1

WHO (1988a) Salmonellosis control: the role of animal and product hygiene. Report of a WHO Expert Committee, Geneva, 22–29 Sept 1987 Tech Rep Ser 774

WHO (1988b) Food irradiation, a technique for preserving and improving the safety of food. World Health Organization, Geneva 354

WHO (1988c) Foodborne listeriosis. Report of a WHO informal working group, Geneva 15–19 February 1988, WHO/EHE/FOS/88.5

WHO (1989) Health surveillance and management procedures for food handling personnel. Tech Rep Ser 785, WHO, Geneva

WHO/FAO/IAEA (1989) International consultative group on food irradiation. Consultation on microbiological criteria for foods to be further processed including by irradiation. WHO/EHE/FOS/89.5 WHO Geneva

Whyte GN (1959) Principles of radiation dosimetry. Wiley, New York

Williams PC, Norris KH (1987) Near infrared technology in agricultural and food industries. American Association of Cereal Chemists, Saint Paul, MN

Willix RLS, Garrison WM (1967) Chemistry of the hydrated electron in oxygen-free solutions of amino acids, peptides and related compounds. Radiat Res 32: 452–457

Woodburn M, Bennion M, Vail GE (1962) Destruction of salmonellae and staphylococci in pre-cooked poultry products by heat treatment before freezing. Food Technol 16: 98–100

Wright RL, Walker HW, Parrish FC Jr (1986) Survival of *Clostridium perfringens* and aerobic bacteria in ground beef patties during microwave and conventional cookery. J Food Prot 49: 203–206

Wu WS, Wu CHH, Fu YK, Chu SL, Tsai CM (1980) Effects of gamma radiation and chemical treatments on the control of postharvest diseases of potatoes and onions. Mem Coll Agric Nat Taiwan Univ 20: 17–24

Yang GC, Mossoba MM, Merin U, Rosenthal I (1987) An EPR study of free radicals generated by gamma-radiation of dried spices and spray-dried fruit powders. J Food Qual 10: 287–294

244 References

Yoshida H, Kajimoto G (1988) Effects of microwave treatment on the trypsin inhibitor and molecular species of triglycerides in soybeans. J Food Sci 53: 1756–1760

Yoshida H, Kajimoto G (1989) Effects of microwave energy on the tocopherols of soybean seeds. J Food Sci 54: 1596–1600

Yoshida H, Hirooka N, Kajimote G (1990) Microwave energy effects on quality of some seed oils. J Food Sci 55: 1412–1416

Young GS, Jolly PG (1990) Microwaves: the potential for use in dairy processing. Aust J Dairy Technol 45: 34–37

Young RE (1965) Effect of ionizing radiation on respiration and ethylene production of avocado fruit. Nature (Lond) 205: 1113–1114

Yousef AE, Marth EH (1985) Degradation of aflatoxin M_1 in milk by ultraviolet energy. J Food Prot 48: 697–698

Zabielski J (1989) Effect of gamma irradiation on the formation of cholesterol oxidation products in meat. Radiat Phys Chem 34: 1023–1026

Zachariah NY, Satterlee LD (1973) Effect of light, pH and buffer strength on the autooxidation of porcine, ovine and bovine myoglobins at freezing temperatures. J Food Sci 38: 418–420

Zegota H (1988) Suitability of Dukat strawberries for studying effects on shelf-life of irradiation combined with cold storage. Z Lebensm Unters Forsch 187: 111–114

Zhao KJ, Wan H (1987) A study of the effects of gamma irradiation on the respiration and ethylene production of apple fruits. Acta Hortic Sin 14: 35–41

Zimmermann WJ (1983a) Evaluation of microwave cooking procedures on ovens for devitalizing trichinae in pork roasts. J Food Sci 48: 856–860

Zimmermann WJ (1983b) An approach to safe microwave cooking of pork roasts containing *Trichinella spiralis*. J Food Sci 48: 1715–1718

Zimmermann WJ (1984) Power and cooking time relationships for devitalization of trichinae in pork roasts cooked in microwave ovens. J Food Sci 49: 824–826

Zimmermann WJ, Beach PJ (1982) Efficiency of microwave cooking for devitalizing trichinae in pork roasts and chops. J Food Prot 45: 405–409